Bioshelters,
Ocean Arks,
City Farming

Bioshelters, Ocean Arks, City Farming

Ecology as the Basis of Design

Nancy Jack Todd
and
John Todd

Sierra Club Books
San Francisco

Copyright © 1984 by Nancy Jack Todd and John Todd

Bioshelters, Ocean Arks, City Farming is produced under the auspices of the Society for the Study of Native Arts and Sciences, a nonprofit educational corporation whose goals are to develop an ecological and crosscultural perspective linking various scientific, social and artistic fields; to nurture a holistic view of arts, sciences, humanities, and healing; and publish and distribute literature on the relationship of mind, body, and nature.

Cover design: Paul Bacon
Book design and graphic production by Paula Morrison

Library of Congress Cataloging in Publication Data

Todd, John, 1939-
    Bioshelters, ocean arks, city farming.

    Bibliography: p.
    Includes index.
    1. Human ecology. 2. Bioengineering. 3. Architecture, Domestic—Environmental aspects. 4. Fish culture. 5. Organic farming. 6. New Alchemy Institute. I. Todd, Nancy. II. Title.
GF50.T6 1984   304.2   83-51436
ISBN 0-87156348-7
ISBN 0-87156814-4 (pbk.)

Printed in the United States of America

10 9 8 7 6 5 4 3 2 1

Table of Contents

This book is dedicated to our parents:
To Mildred Ruth Todd, Robert Thomas Todd,
Mary Evelyn Jack and to the memory of
Vaughn Ainsworth Jack.

Illustrations by **Jeffrey Parkin**

With contributions from
        Keith Critchlow (Lindisfarne Chapel)
        David Sellers (Cathedral Bioshelter)
        Paolo Soleri (Arcosanti)
        Christopher Swan (Sun Train – Light Rail)
        Jeff Zwinakis (Ocean Arks and Ocean Pickups)

Acknowledgements

As is obvious from the opening pages, the ideas presented in this book are based on the work of a number of people and, because of this, we see them as having contributed to it. Foremost among them are our friends and co-workers at New Alchemy. As we have said before, their dedication is the real alchemy and to them we are and shall continue to be indebted. We also mention the happy meeting of minds that took place when we became involved with William Irwin Thompson and his colleagues at Lindisfarne. Of our friends there Drs. Lynn Margulis and James Lovelock, Sim Van der Ryn, Paolo Soleri, and Hunter and Amory Lovins have had a direct and important effect on our work. Architect Malcolm Wells has been an influential voice in persuading us that the design ideas that had arisen in the context of a solar village were universally applicable, and particularly so in an urban context. The Very Reverend James Parks Morton and Pamela Morton of the Cathedral Church of St. John the Divine in New York City have been invaluable allies in helping to dispense our ideas on ecological design and to embody them within the context of traditional Christianity.

Our ability to express our ideas would have been seriously curtailed without those who helped us by portraying them visually. To Jeffrey Parkin as the principal illustrator we owe the translation of biological concepts into applied forms. The talent for conveying transition and succession, being and becoming, is a rare one. Jeff Zwinakis, Paoli Soleri, and David Sellers and his associates have been extremely generous with their illustrations. We are equally grateful to Dan Blackwood for his careful photography of many of the original drawings. To navy architects Richard Newick and Philip S. Bolger we extend our thanks for lending their talents to our waterborn projects.

In working directly on the book we wish to express enormous gratitude to Lindy Hough and Richard Grossinger. Lindy has been a creative, exacting, and superb editor. In addition to the work involved in producing the book, Richard never allowed any of us to forget the larger vision that sustains it. Their hard work, faith in the ideas, and friendship have been instrumental to us.

Heartfelt gratitude to the person who typed one's manuscript is a traditional part of almost every author's roster of acknowledgements. In our opinion, however, few typists have been the recipients of pages as ill-typed, corrected, revised, and rearranged as Cynthia Knapp accepted from us. Beyond translating occasionally close to indecipherable markings into immaculately typed manuscript, she gently questioned inconsistencies and redundancies, and generally made us respectable. More remarkable perhaps, she never lost her temper and has remained a valued friend throughout.

We are also extremely appreciative to our other friends and to the members of the Summer Dance Theater who not only have been wonderfully supportive but from time to time have provided enough much needed distraction to bring us back restored to the task at hand. The contribution of Sherry Greene Starr at a particularly critical time was as moving as it was essential.

Finally, to our children and our extended family who have been a part of the household during work in progress, our thanks for being as quiet as the innate volatility of most of us has allowed. While the silence may occasionally have fallen short of the sepulchral, the understanding and enthusiasm for our work and the love with which we are surrounded has been and always will be the sustaining force.

The rule of no realm is mine, but all worthy things
that are in peril as the world now stands, those are my care.
And for my part, I shall not wholly fail in my task
if anything passes through this night that can still
grow fair or bear fruit and flower again in days to come.

J.R.R. Tolkein

Chapter One

New Alchemy:
Where It All Began

Suppose there were a clever global pollster, assigned to travel the world, questioning people in places as disparate as the far reaches of the Australian outback, or in downtown Detroit, on one of the Greek islands possibly, or in Brazil or on a rural Chinese commune. Suppose the object of the study was to learn what, if anything, is universal to the human experience—what it is that matters to people ultimately. Would there be any commonality of response? Would it be that, after relief from deprivation and suffering, most of us would name love, concern for our children and families, hopefulness, peace of the heart? Or would it more likely be possessions, wealth, power? It seems possible, were we to answer truly, that we would most want those intangibles without which all other achievements eventually prove barren. Yet, as world events attest daily, as a species we behave as though power and material objects alone were worth our striving.

This is a book about ecological design. By this term we mean design for human settlements that incorporates principles inherent in the natural world in order to sustain human populations over a long span of time. This design adapts the wisdom and strategies of the natural world to human problems. Implicit in this study there is a larger question—what is the role of humanity in the greater destiny of the Earth? As scientific research continues to discover, all of us who inhabit this planet share the same kind of genetic material. In terms of biochemical make-up and genetic structures, the similarities between the human being and the bacteria, for example, are greater than the differences. The illusive and pervasive issues of how human beings, as the only self-conscious species, are to live in the world is a logical outgrowth of our new biological knowledge. Even if the present path of industrial society held much promise of survival, which we feel it does not, it is a violent and unhappy world. A reevaluation of the way humans place themselves in the larger world seems timely, if not overdue.

In recent years people everywhere have been experiencing a reawakening realization of the Earth as a planet, alive and beautiful beyond words. Photographs from space have affirmed its incandescent uniqueness. Scientists, ecologists and environmentalists are steadily increasing our knowledge of its complexity and vulnerability and are rapidly restructuring our understanding of it. Over much the same period our own research in applied biology and biotechnology has led to an emerging synthesis of precepts by which the present human community could sustain itself indefinitely without destroying its basis of living support systems. It is a claim, we think, that could not be made for current industrial cultures. Co-evolutionary with a reawakening sensitivity to the life of the planet there has developed a series of insights, methodologies, and technologies that make it possible to create a post- or meta-industrial society without violating fundamental ecological integrity. This ability is as unprecedented as it is timely.

New Alchemy

Our own work in ecological design began many years ago with the New Alchemy Institute, a research institute on Cape Cod, which we founded with the aquatic biologist and writer William O. McLarney in 1969. Although it is not within the province of this book to recount in detail the terryifying and all too probable threats that imperiled the world then, and more so today, the litany of woes, headed by nuclear war and widespread ecological disaster that prompted us to take some sort of action at that time remains long and familiar. Barring cataclysmic events, we confront daily an unrelenting array of humanly created problems. The arms race, local war, industrial and agricultural pollution, contamination of air, land, and water, nuclear meltdown, acid rain, deforestation, desertification, famine and homelessness are far from ephemeral spectres. They are directly and achingly real to many people in many parts of the world and indirectly affect everyone. Yet it is the same haunting threat of rapid or gradual extinction of much of life on Earth, at least in its larger forms, that has been a catalyst and context for many changes, some of them hopeful, which are beginning to take place. This threat was certainly instrumental to the birth of New Alchemy, and for the thinking and the work that we shall be discussing throughout the book.

In 1969, compounding the ongoing crises of the Vietnamese War and conflicting social issues of the late sixties, there was an unremitting flow of information about the destruction of the environment. The word ecology, which, like economy, is derived from the Greek *OIKOS*, meaning household or home, was being adapted into the general vocabulary. Unlike the term *environment*, which denotes one's surroundings in an objectified sense, *ecology* by its very inclusiveness implies interconnectedness. This word's incorporation into daily language was some indication of a dawning realization of the complexity of the human interaction with the natural world.

About that same time we were fresh from a more directly experiential event that also had shaken us and given us a realistic sense of our own capabilities. Friends renting a small ranch in the hilly country west of San Diego in California and just north of the Mexican border had asked us for help in studying the ranch so that they could grow toward self-sufficiency in food and energy. Spending time in the dry, sun-drenched hills there, we came to realize that, given the lack of water and almost non-existent soils where only manzanita bushes seemed to flourish in quantity, we had no idea how our friends should proceed. Academic credentials and training in biology were no help in redefining the human place in what had initially seemed a hospitable landscape, at least not within the framework of the ecological ethic. We decided then to undertake an intensive study of the land to examine as many of the aspects of the environment as we were able to. We collected, studied, and catalogued soil samples, soil animals, insects, plants, shrubs, and rocks, and noted trees, birds, animals, and other fauna. Very slowly, some relevant clues begam to emerge. Midway up a small gorge we found a plant, the roots of which are known to seek out moisture, indicating the presence of a hidden spring. Below, where the gorge began to flatten out, there was a live oak tree surrounded by an association of plants that included miner's lettuce which we learned required good soil. With this discovery of a source of water and suitable soil, establishing a garden and a move toward food production became a possibility. If the spring were to be tapped as a source for irrigation and fish ponds, then gardens, poultry and other livestock, and eventually orchards could be integrated and an agricultural ecosystem could become an achievable goal. Our hopeful prognosis for the ranch had an abrupt but unhappy end, however, for the rent on the land was raised unexpectedly to well beyond the means of our friends. Shortly afterwards bulldozers appeared on the crest of the hill and began levelling for a colony of weekend cottages.

We emerged from the experience realizing the need for a more secure basis for research. This combined with a desire for the flexibility and freedom to begin to search for relevant knowledge, new and lost, led us outside established academic and scientific institutions to consider an organization of ur own. Our initial impulse was to become disseminators of information, thinking, naively, that this would trigger the necessary reform. We became involved in communication with other like-minded people—biologists, naturalists, students, environmentalists, parents, anyone similarly concerned. But even as we decried, privately and publicly, the wastefulness and destructiveness of many of the practices of industrial societies we had little to suggest in the way of other possibilities. We began to ponder the possibilities for ecological analogues to the current, biologically insupportable, industrial methods for sustaining the people of the world. We wondered whether humanity could ever hope to exist again in a mutually supportive and beneficial way within the biosphere. This question, once it had articulated itself, persisted. It gnawed at our minds and expanded to become the underlying intellectual paradigm on which we were to found New Alchemy.

Accepting the likelihood that there were no existing institutes that would allow us the freedom of crossing disciplines, setting different values and priorities as the basis of our work, and looking at biology and agriculture in a larger social and cultural context, we created our own fledgling institute, adopting as we did so a credo that may have appeared pretentious or absurdly quixotic. It was and remains utterly heartfelt. Our logo read:

"The New Alchemy Institute.
To restore the land, protect the seas,
and inform the Earth's stewards."

Our approach to so large and amorphous a mandate was to translate it into research which would show how to affect a shift to a more ecological basis for the provision of basic human needs. This was the work we began when we, and soon afterwards Bill McLarney, crossed the country and rented a twelve-acre farm in Massachusetts. New Alchemy settled in on its Cape Cod center in late 1971.

There were only two paid staff members who did the administrative work when we first moved onto the farm. Everyone else then, the half dozen regulars we considered staff, including Bill McLarney and ourselves, had jobs at The Woods Hole Oceanographic Institute or elsewhere. The rest of the work and the maintenance was done by volunteers and friends who gave what time they could. Perhaps by dint of the fact that it was exploring the potential of a new paradigm, New Alchemy, from the beginning, was difficult to fit into existing funding structures. It remains so today, with a core staff of twenty-two and a long record as an innovative research and education organization. Although several of the staff commanded large grants when they worked in established academic and research institutions, New Alchemy consistently has had to struggle to stay alive financially. Over the years the work has been, and continues to be, carried on by staff who may see months go by between pay checks, and by apprentices and volunteers.

To investigate a question as large as an alternative means for providing basic human needs, the research was divided into the tangible and more approachable areas of food, energy, and shelter. Starting quite literally, from the ground up, we discovered that to work in food production we had first to manufacture topsoil to augment the meagre layer already existing on the sandy, glacial terminal morrain which forms Cape Cod. As we did so, we planted the organic vegetable gardens that have been the focus of many years of experiments. Subsequently, the agriculture program has expanded into research in many other areas including tree crops, a type of farming that is a logical adaptation to the natural state of Cape Cod. Young orchards were started, as well as stands of trees for lumber, fuel, manufacture, and food for livestock.

The other major branch of our research in intensive food production was in the field of aquaculture, the culture of fish and aquatic animals. Aware of the growing protein defienciency in the diets of people in so much of the world, we began research into methods of producing protein resources that were both economically accessible and ecologically benign. In investigating possible alternatives to nuclear and fossil fuel energy, our work has been mainly with the renewable sources of the sun and wind. Windmills, looking, as one of the group put it, "like steeples of a solar age," and small, sun-trapping domes have dotted the New Alchemy landscape since its first summer. Our efforts proved encouraging in both agriculture and aquaculture within a relatively short time. We began to see a great improvement in the soil and good-to-excellent yields from the gardens and the sun and wind-powered fish ponds. We were emboldened to go on to the issue of shelter—to try to create an integrative form of architecture that would incorporate renewable energies and biological systems in the form of growing areas for plants and fish.

From early on our work found unique direction in this harnessing of wind and solar energy to power biological systems. We built a number of variations of small translucent structures that were both greenhouse and aquaculture facilities. They were microcosms that absorbed and intensified the pulses of natural forces to provide food and an optimal environment for life forms ranging from soil animals, to plants, to fish, to people. As one design improved on another we evolved what was named the bioshelter, the structure at the core of much of the achievement of the Institute during its first decade. Beyond the Institute, it has been a major catalyst in exploring the fruitfulness of a marriage between biology and architecture. It is at the core of most of the design concepts to be described throughout this book—a harbinger for new directions in public buildings, commercial greenhouses, private and aggregate housing, and year-round community gardens.

Whether it was timeliness or Fortune that smiled on us, New Alchemy's work was recognized in a shorter time and on a larger scale than we would have conceived possible in our first seasons. By 1976 we had designed and built two large bioshelters.[1] One was on Prince Edward Island in Canada, established in cooperation with Canadian Federal and Provincial authorities, as a part of conserver society policies. The other was on our farm on the Cape and designed in collaboration with Solsearch Architects. Although bioshelter has become a generic name for such structures, we chose, more poetically, to call them respectively the Prince Edward Island and Cape Cod Arks. As the cell is acknowledged as the basic building block in organic evolution, the bioshelter is likely one day to be seen as a basic building block in ecological design. The Ark in Canada was opened with considerable fanfare, attended by Prime Minister Trudeau and the then Premier of Prince Edward Island, Alex Campbell. The building remains well ahead of its time as an experiment in systemic design and in the incorporation of biological elements into a structure that was greenhouse, aquaculture facility, and residence for the people who worked in it. Seen in retrospect, what we were acknowledging at the opening of the Ark in Canada was a turning point in integrative architecture. After several years of our own research there, monitoring and testing the building, we turned it over to provincial authorities and it is now a commercial trout-raising facility and hatchery.

After 1976, a great deal of research was done in both Arks on interior climate, energy requirements, soil, and vegetable and fish production as well as overall performance. In the Cape Cod Ark an elaborate monitoring system was installed with seventy-six sensors relaying information on the ongoing state of the building to a central computer. Both it and its Prince Edward Island counterpart weathered their first, unusually severe winters, without resorting to other than solar heat except for a woodstove in the living area of the Canadian Ark. Assessing and extrapolating the ongoing per-

The Cape Cod Ark

formance and productivity of the buildings, we were able to pronounce the Arks viable beyond our early hopes. They were independent of outside energy sources for heating and cooling and yielded well throughout the year, portending an economic base for future replicas.

For almost all of the same period New Alchemy concepts were being researched and tested simultaneously in Costa Rica. In 1973 Bill McLarney founded NAISA (New Alchemy Institute Sociedad Anonima). It consists of a small farm located in the Atlantic lowlands, on the Caribbean coast, just north of the Panamanian border. NAISA is an integral part of the Gandoca community there. While maintaining a lifestyle absolutely consistent with that of his neighbors, Bill has been able to bring considerable financial aid and technical innovation into the area. For all that, NAISA remains a Costa Rican organization with in-country directors and local staff and apprentices. Its principal mission has been to integrate locally defined rural development and ecological conservation, which involves work in aquaculture, agricultural crop diversification, and local economic development. Although NAISA is completely independent, there is considerable exchange between it and New Alchemy on the Cape in terms of staff, apprentices, and information.

On Cape Cod the Institute has expanded until now, with the exception of the small public library and reading rooms, every inch of the farm-house is crammed with desks and offices. The old dairy barn houses not only a workshop and storage areas, but a lab, a computer facility, and a newly completed super-insulated energy education auditorium which demonstrates the most advanced materials and concepts in conservation. From a starting point behind the house, a series of signs steers ten thousand visitors a year through a self-guided tour of the farm. The education and outreach program offers a number of guided tours as well. Beyond the lawn and the row of experimental bioshelters which dot the hill overlooking the garden sits an innovative pillow dome (so named because its translucent skin is divided into hundreds of tiny pillows) that was honored at its opening in June of 1982 by the presence and approval of the late Buckminster Fuller.

Although the bioshelter research, because of its contained nature, lends itself well to description, the agricultural work on the farm also produced encouraging results. Yields of organic vegetables on the steadily improving soil tripled Department of Agriculture averages. Some of the aquaculture in solar driven tanks rivals world production records. Farther back from the gardens, and away from the house, are extensive herb gardens and a growing orchard of young fruit and nut trees continually tested for adaptability to the Cape's climate and soil. The experiment continues behind the barn where many more trees, food-producing and nitrogen-fixing shrubs, and many different species of bamboo, are all patrolled regularly by a vociferous gaggle of weeding geese.

Although the present Institute has been pronounced by U.S. Senator Paul Tsongas as "an island of reason is what otherwise is a rather dismal scene in this country," a visitor arriving at the farm some years from now is likely to see as great a difference again as one from the early years would see now. Just as then a few of us then could envision not just a neglected farm on a wooded oasis in the midst of suburbia, but a synthesis of ecological design through faithfulness to the integrity of the natural world, present New Alchemists see beyond the Institute of today to the one it can become. Then only renewable energy will be used as a source of power. The trees will have grown to yield fruits, nuts, animal feeds, fuel, and building materials. The forest stands, like the fish, fresh and smoked, and the produce from the gardens and the bioshelters will find ready local markets and be a major source of economic support for the Institute and its staff. The farm will have become a model of what is possible in the wooded northeast, its ecosystem having proceeded through the phases of ecological succession to a near climax state where the soil and plant communities are renewed and maintained in an ongoing productive yet stable condition. As our farm in the image of the forest reveals a pattern of sustainability for our own bioregion, the hope is that the same kind of thinking and observations can be applied to other areas to create farms in the image of the prairie, the desert, or the savannah.

By the end of its first decade, pooling the promising results of New Alchemy's research with that of others in related fields, we began to feel that we had achieved an affirmative answer to our original question. An alternative, ecological human support base was feasible[2]. It was a sound idea to look to biology as a basis of design. We see this as an informed affirmation of the regenerative capabilities of the planet and of the human role as stewards of the Earth. For those of us who accept its validity, it engenders a new and hopeful way of looking at the world. Unlike present industrial practices, it is reversible. We can afford mistakes. Failures can be recycled into more useful forms and tried again, leaving open the possibilities for continual choice. The thinking that underlies the ecological paradigm is less a linear Cartesian model but rather of the mode that can best be envisioned by the hologram, embodying ceaseless mutual causality and interdependence. If information can be defined as a difference that makes a difference, the work done at New Alchemy can be considered to have helped to create a new piece of information, a variable in the overall fund of human knowledge. It represents a new way of knowing. It has reinforced our own conviction that a smoggy guttering out of life is not inevitable, but becomes, through non-action, a choice. This conviction has found strength in the results of the work of countless other like-minded groups around the world in agriculture, ecology, cybernetics, materials science, physics, and humanistic psychology, economics and politics. Should the ecological paradigm become, as it could, a governing world view and trigger adaptive behavior in significant numbers of people, we would find ourselves with a renewed promise for the future.

Chapter Two

From Bioshelters to Solar Villages To Future Human Settlements

The years with New Alchemy, culminating in the design and building of the Arks, laid the foundation for our work in ecological design. It became rounded out for us intellectually as word of our research and its implications for the workability of an ecological paradigm began to spread. New Alchemy's credibility became established. We met a number of people who, through their writings, had been our mentors when we had been establishing the underlying intellectual framework of the Institute. Among them were the economist E.F. Schumacher, author of *Small is Beautiful*, historian and social critic Murray Bookchin, cultural historian William Irwin Thompson, *Whole Earth Catalog* and *Co-Evolution Quarterly* editor Stewart Brand, Buckminster Fuller, anthropologist/ecologist Gregory Bateson, and anthropologist Margaret Mead. From all of them we received approval and encouragement. Gregory Bateson made what was to us the most memorable and heartening statement of reinforcement when he pronounced what New Alchemy stood for–"an epistemology with a future."

For several years during this period in the mid-seventies, William Irwin Thompson had been urging people working in futuristic concepts to think in larger terms and to reconsider the nature of human settlement. He advocated a post- or meta-industrial village, which he called the *deme*, as the next unit for design. It was Margaret Mead, however, who was the most immediate catalyst for the next stage in our work. Shortly before her death she sent for us to explain that she thought that the time had come for New Alchemy's work to be applied on a much broader scale. Her message, in essence, was that the creation of the bioshelter had been a good piece of work, useful and relevant, but at the level of the private structure or single family house it was limited. Most of the people in the world would neither be able to afford one nor have access to one. She felt we must begin to envision the same kind of integrative architecture at the level of the block, the neighborhood, or the village.

It was a challenging legacy. We could not, nor would we, have refused. It was a major next step conceptually–the question of how to take the idea of using biology as the model for design, which had proved workable on a small scale in an experimental setting, and begin to apply it on a much broader scale. Our first response was to ask people from a range of disciplines to join us and the New Alchemists to meet together and ponder the question for several days. In April of the year following Dr. Mead's death we convened a conference entitled, *The Village as Solar Ecology: A Generic Design Conference*[1]. We dedicated the conference and the report resulting from it to her memory.

The working hypothesis for the meetings was given in the original proposal which stated:

> "The blending of architecture, solar, wind, biological and electronic technologies with housing, food production, and waste utilization within an ecological and cultural context will be the basis of creating a new design science for the post-petroleum era."

We were not so naive as to expect that we would emerge from the conference complete with a communicable and tangible epistemology for the design of sustainable communities of the future. To have done so would have been beyond the scope of any conference that did not go on for years. At that conference, however, and in subsequent years, a number of design precepts have emerged from our own and other organizations which are fundamental to the planning of existing and new communities. If we are to continue to shelter and feed the people of the world in the coming centuries, we will have to design in a different way than we do now. Gradually, these precepts are being incorporated into existing or planned projects. In 1982 we took a further step toward implementing the application of biotechnology when we created Ocean Arks International. It is a non-profit organization intended to disseminate the ideas and practice of biotechnology and ecological sustainability throughout the world. Its first project has been to design and build a high speed sailing vessel, the Ocean Pickup, for use by fishermen.

That a revisioning of the way we live and think about the planet is crucial was reinforced unexpectedly for us not long after the Village Conference when a young woman from the Wampanoag tribe at Mashpee, Massachusetts visited us at New Alchemy. As always, when the weather is even faintly kind, we were sitting outside on the grass to talk. The general subject of our conversation was the differences in the ways of our respective cultures–hers, the ancestral traditions of the Wampanoags and ours, at least in her eyes, those of an exploitative technological society. What she said to us in essence was, "My people don't understand you or why you do the things you do. We don't understand why you are still trying to take more of our land. Why must you own things. Why must you always have more." Her eyes clouded for a moment as she searched for the right explanation, then she gestured to a neaby flower bed. "A seed, a flower, a tree unfolds according to the instructions it has been given. We have always tried to live by ours. We don't understand yours. How you have been taught to live. What your instructions are."

It is possible that we have forgotten. What *are* our instructions, as a people, a culture? The observation of our Wampanoag friend was strangely evocative of the words of the poet Annie Dillard, when she says in "Teaching a Stone to Talk," "I want to learn or remember, how to live."[2] At the Solar Village Conference a few years earlier, architect and Royal College of Art Professor in London Keith Critchlow had spoken to us of ancient concepts of design based on a sacred geometry of the Earth itself. The great circle arches, and the mathematical laws that govern the movements of the stars, have found expression in form structures, building and culture. These are the expression of the psyche of a people when the sacred is an underlying energizing force. Carl Gustav Jung spoke of such a time as "a period in which man was still linked by myth with the world of the ancestors and thus with nature truly experienced."[3] It is a world view long gone from us now, but unlike our own governing mythologies of progress and material happiness, it held and satisfied minds and souls for untold ages and generations. As Giorgio de Santillana said in *Hamlet's Mill*, "...it lived on and flowered and let the world live."[4] For longer than we can know, humanity lived in a universe governed by fixed laws at once mysterious and predictable, the visible aspects of which were inhabited by innumerable unseen gods, spirits, and forces. Spirit and matter, humanity and nature were one, an original seamless, undivided dynamic unity that encompassed and enclosed the interplay of all forces.

But then a new cosmology became dominant in the Western world. It has taken place in a brief period, in comparison to the preceeding eons.

This new cosmology began with some of the philosophers of ancient Greece and evolved over the centuries to maturity in Europe during the Age of Enlightenment. What Louis Mumford has called the mechanical world order came to be increasingly accepted. Reflecting the thinking, among others, of Johannes Kepler who, in 1605, wrote "My aim is to show that the celestial machine be likened not to a 'divine organism' but rather to a clockwork," it became the basis of all scientific exploration. That knowledge which could be measured, quantified, charted, and ultimately objectified was seen as legitimate for study. Basing their thinking on the philosophy of René Descartes, who saw a fundamental division between the separate and independent realms of mind and matter, scientists began to treat matter, including the living world, as lifeless and apart from themselves, leading to both the greatness and the folly of the idea of scientific objectivity. The influence of Cartesian duality was to further lead Western cultures to think of the mind as divided from the body and from the unconscious mind. It remained for Isaac Newton to complete the intellectual domination of the mechanistic world view, by using it as the groundwork foundation for his construction of classical physics. The philosophy and scientific discovery of the Age of Enlightenment produced an explosion in knowledge and provided the basis for the ever-accelerating explorations of science. That science bred the technology that has changed the face of the Earth. Because that technology could not have been implemented without a world view that reflects our fundamental attitudes about life and mechanism, we are at once its benefactors and its victims.

For a long time, certainly into our own childhood, the images of a clockwork universe and of the natural world as functioning mechanistically, machine-like, removed and separate from ourselves held sway. It is still not broadly questioned, but signs of discomfort with it are becoming increasingly common. Feedback from the environment—scarred, denuded hillsides, dying lakes, air murky on the horizon, miles of scrapheaps and automobile graveyards, urban sprawl—has become impossible to ignore. The evidence has gained sufficient proportions that it does not take a philosopher or a specialist of any kind to wonder whether our treatment of the natural world as immune to abuse is the best or even a safe course. Gregory Bateson compared our behavior to the situation in Lewis Carroll's *Alice in Wonderland* where Alice found herself obliged to play croquet with a flamingo for a mallet and a hedgehog for a ball. Neither the flamingo nor the hedgehog behaved in a predictable manner, because they were alive, not lifeless and inert tools, with Alice the only living variable involved in the interaction. Dr. Bateson terms it "an inappropriate coupling of biological systems," implying, obviously, that we treat other living systems as inert tools in our mechanical model of the universe. This led him further to question whether information after being processed through the conscious mind is adequate for understanding another biological system, the behavior of which is based on complicated patterning on a non-conscious level.

We have not been inclined to think a great deal about the assumptions implicit to the mechanocentric theory of the universe, the mindset of scientists being no less immune to the pervading climate of thought than that of the rest of us. We have all been nurtured in the unconscious acceptance of the concept of progress and put our trust in linear modes of thought and causality. It is a habit of mind that has become second nature and we are reluctant to trust any other. Alfred North Whitehead observed, "Our science has been founded on simple location and misplaced concreteness," and "science divides the seamless coat—or to change the metaphor into a happier form, it examines the coat, which is superficial, and neglects the body, which is fundamental. The disastrous separation of body and mind which has been fixed on European thought by Descartes is responsible for this blindness of science."[8]

For some time now there have been tremors threatening to undermine the edifice of Descartes and Newton and the acceptance of the strictly causal nature of physical phenomena. Early in this century the general relativity theory of Albert Einstein abolished the concept of absolute space and time, and with it the mechanistic world view. Since that time the cosmology on which we have built our political, economic and social structures has no longer fit the theories being involved on the frontiers of scientific theory and advanced thought. In spite of the daily evidence of our senses and the quantity of information pouring in from the media, books, and scientific journals—deserts expanding, forests dwindling, species vanishing at the rate of one a day, widespread social unrest—we continue to act as though none of this has anything to do with us and our behavior. Reports are published on limits to growth, on the finite carrying capacity of the Earth, on repression and injustice, yet economic and political strategies, both capitalist and communist, continue to be based on assumptions of indefinite exploitation and continued growth. This reflects our world view which was built on now outdated concepts but is no longer cohesive with emerging scientific thought. Gradually, however, as general loss of faith and confidence are becoming evident, this paradigm is fading and another is emerging. As the concept of a mechanistic universe and a schizophrenic attitude to nature are relinquished, we find ourselves on the verge of a cosmology potentially far more cohesive intellectually, more sound intuitively, and more peaceful spiritually.

Such a realization led Murray Bookchin to state in the introduction to *The Ecology of Freedom*, "Such a philosophy has always been more than an outlook or a mere method for dealing with reality. It has also been what philosophers call an ontology—a description of reality conceived not as mere matter but as active, self-organizing substance with a striving toward consciousness."[9] The British physicist David Bohm has further stated that "…inseparable quantum connectedness of the whole universe is the fundamental reality."[10]

It begins to become increasingly apparent then that with the slow dissolving of the mechanistic world view, we are evolving a new or renewed awareness of the universe—one that is internally consistent. No longer must we gloss over the discrepancies between the spiritual and the material, the sacred and the secular. The scientific paradigm points to acceptance of the cosmic dance of Shiva and the dance of quanta—and all of us participate, creatures of light-energy, star matter; all are dancers. The ancients watched the sky and saw with their hearts, Eastern mystics see with the inner eye, and now physicists have looked at the universe with telescope and microscope and all seem to have come to a commonality of understanding. The physicist John Wheeler maintained that the most important aspect of the quantum principle is that it destroyed the concept of the world as "sitting out there." The act of observation in itself makes the observer a participator. Making a measurement, even of an electron, changes the state of the electron to that degree so the universe will never again be quite the same. He concluded, "In some strange sense the universe is a participatory universe."[11]

The idea is an overwhelming one. Just as we are beginning to reassume some responsibility for our actions in the context of the Earth, a question of a larger context and consequence arises. Yet no matter how far reaching our ultimate accountability, it seems common sense that it is here, with the Earth and each other, that the healing must begin. And if because we have discarded their myths we can no longer look to the ancient gods for instructions as to how to proceed, science reembedded in the cosmology of a participatory universe and a sense of the sacred may yet prove to be an appropriate guide. Writing in his *Notebooks*, Charles Darwin stated, "The grand question which every naturalist ought to have before him, when dissecting a whale or classifying a mite, a fungus, or an infusorian is 'What are the Laws of Life?'[12] More than a hundred years later and, by his own admission, not much farther along with so central an issue, Gregory Bateson, in *Mind and Nature*, asked the same question in different words: "What pattern connects the crab to the lobster and the orchid to the primrose and all the four of them to me? And me to you? And all the six of us to the amoeba in one direction and to the backward schizophrenic in another? What is in the pattern," he goes on to ask, "which connects all living creatures?...the pattern of patterns?...the metapattern?"[13] And he added that, as of 1978 when he wrote the book, there was no conventional way of even beginning to describe the tangle of the vast network of interrelationships and our ideas about them which would be necessary to grapple with the metapattern. An intuitive recognition of such a metapattern, however rudimentary, was fundamental to evolving the underlying paradigm for New Alchemy and for all the biological design work there.

The chance to participate in a process so much larger than ourselves holds out to us, the heirs of the age of science and technology, the possibility of a new set of instructions—or perhaps the eternal instructions—in a language, that of science, which we understand and accept. The Native American spokesman, Chief Black Elk, once declared that "All life is holy and good to tell" but we chose not to listen.[14] Yet we cannot ignore the rest of life indefinitely. Our understanding must grow to encompass a union of nature and culture in which the sacredness of all life is honored. As long as we saw all other life as outside and apart from ourselves, we treated it carelessly. Embracing the interconnectedness of all life, we can again weave together the rift between sacred and secular, and the totality will be seen as sacred. Perhaps, now, with a synthesis of knowledge of fields as disparate as quantum physics, astronomy, ecology, religion, holography, anthropology, and the contemplation of sacred art, architecture, and geometry, certain harmonies are beginning to be heard, or heard again, and our sense of the world, rather than being cacaphonous and diffuse with the claims of economists and environmentalists, communists and capitalists, the secular and the sacred, begins to make more sense, to ring true. Perhaps a cosmology that is at once beyond memory and still just out of reach of present knowledge, yet somehow alive within us, is unfolding. The stars are still there to remind us that we are both trivial and non-trivial. One way of reaching out toward what we want to bring into being is a careful reassessment of how we are to live; where under the shining sky, in what relation to the sun and the solar winds, and how we are to best care for the living, celestial matter that is the small portion of the Earth on which we find ourselves.

Chapter Three:

Emerging Precepts of Biological Design

From the kinds of work, from the experimentation and the observations of our years at New Alchemy, we began to evolve a way of looking at and thinking about the world, an epistemology, to use one of Gregory Bateson's favorite words. As we began to apply this type of mindset, initially to problems at the Institute and subsequently, on a broader scale in many parts of the world, it became evident to us that we were creating a series of precepts for biological design that could serve both to teach such concepts and to make them replicable in different settings. The articulation of these guidelines for design grew from the confluence of New Alchemy's work with that of a number of other people who had been thinking along similar lines. The formulation of these early precepts as they are applied and tested will contribute, in time, to the creation of a science of applied biotechnology which will serve in turn as a foundation for future design.

Precept One
The Living World is the Matrix for All Design

Although the myriad life forms, processes and natural cycles of the Earth have been thoroughly studied and documented, the question of a pattern of patterns, or a metapattern, that would make the entirety of life comprehensible continues to elude us. The most far-reaching, yet credible theory to date, to our way of thinking, comes from the brilliant research of Drs. Lynn Margulis of Boston University and James Lovelock of England. Called the Gaia hypothesis, after the Greek goddess of the Earth, it suggests that the Earth together with its surrounding atmosphere constitutes a continuum, an entity which, taken as a whole, exhibits many of the properties of life.[1] There it hangs in the blackness of space, like a great, luminous, pulsating cell in and on which, in Dr. Lovelock's words:

> The entire range of living matter on Earth from whales to viruses and from oaks to algae could be regarded as constituting a single living entity capable of maintaining the Earth's atmosphere to suit its overall needs and endowed with faculties and powers far beyond those of its constituent parts.[2]

Formally defined, Gaia may be considered, again in Dr. Lovelock's words:

> ... as a complex entity involving the Earth's biosphere, atmosphere, oceans, and soil; the totality constituting a feedback of cybernetic systems which seeks an optimal physical and chemical environment for life on this planet.[3]

The maintenance of relatively constant conditions by interacting, active control processes may be conveniently described by the term "homeoostatis"–meaning those coordinated physiological processes which maintain most of the steady states in a living organism. The Gaia hypothesis sees the Earth as maintaining homeostatic conditions with the biota actively seeking to keep the environment optimal for life. Dr. Lovelock postulates:

> ... the physical and chemical condition of the surface of the Earth, of the atmosphere, and of the oceans has been and is actively made fit and comfortable by the presence of life itself. This is in contrast to the conventional wisdom which held that life adapted to the planetary conditions as it and they evolved their separate ways.

Such a concept, whether we choose to regard it as scientific fact or as metaphor, is arresting to the modern mind, schooled to think of itself as apart from process and the organic workings of the natural world. The thought that the world around us is alive and continuous with us and through us in ways far more profound that we can know, comes as something of a shock to urban-bred cultures. Whether the Gaia hypothesis is eventually borne out in all aspects, its message is best summarized in a phrase of New Age teacher and visionary David Spangler, "We live in Being." Or, as Stewart Brand once paraphrased Bob Dylan, "In Gaia we're all 'tangled up in blue.'" Trying to fathom and adjust to such a concept is a little like waking from a dream to find that the dream is true. This living entity, made up of billions of interlocking, mutually interdependent, non-human lives surrounds us, contains us, and yet is one with us. Such evidence of a metapattern confronts the human intelligence with questions of a larger intelligence or mind. Gaia is hypothesized as being a cybernetic system. Cybernetic systems are considered intelligent to the extent that they can give a correct answer to at least one question. Dr. Lovelock's most striking example of Gaia's cybernetic capability for self-correction is that of the oxygen content of the atmosphere, which is twenty-one percent, the safe upper limit for life. He states:

> The range of atmospheric oxygen over which fires can take place yet not be so devastating as to threaten all standing vegetation is fifteen to twenty-five percent by volume. It is therefore tolerably certain that atmospheric oxygen has never ranged beyond these bounds in the last several hundred millions of years. This is a truly remarkable feat of regulation, for in the previous ninety percent of the Earth's history the pE has risen by at least ten units but is now held precisely constant.[5]

Such self regulation is protection against the twenty-five percent increase in the sun's luminosity since the Earth had its beginnings. The oxygen content of the atmosphere, it seems, is regulated so as to be optimal for the Gaian entity, for life on Earth as a whole. Furthermore, analysis of the carbon cycles, including atmospheric carbon dioxide and nitrogen cycles, also reveals concurrent, interlocking examples of self-regulation. There is a meta-ecology to the cycles that is extraordinary. These are just two illustrations of the complexity and durability, ingenuity, and elegance characteristic of Gaia, the Earth, co-evolving with and inseparable from life. As Gregory Bateson remarked, "Insofar as we are a mental process, to that same extent we must expect the natural world to show similar characteristics of mentality."[6]

Ironically and surprisingly, although it is more than ten years since the Gaia hypothesis was first published, it seems not yet, as Dr. Lovelock himself ruefully observed, to have set orthodox science on fire. He speculates that perhaps this is because while science swallows the intricacies of relativity and of genetics, it has never been comfortable with whole systems. He feels the circular and recursive logic of whole systems is alien to most scientists. Stewart Brand in the Summer 1983 issue of the *CoEvolution Quarterly* asked editorially, "Why should people who are really worried about non-renewable resources and irreversible damage to the environment take so little notice of a well thought-out optimistic message?"[7] We suspect that it is a matter of mental habit, and of time. The existing world view has been in place for a long time and is only slowly beginning to change. The importance of the Gaia hypothesis to a science of design lies not as a precise tool, or a blueprint, but as a profound multidimensional paradigm for the designs, a meta-model, a basis for thinking about how the world works within which to frame more concrete questions about design.

The relevance of the Gaia hypothesis in the context of our work is as both working premise and metaphor. It is a fundamental premise that all our design must be understood as having as its matrix a living entity that is profound and complex beyond present comprehension. It is a metaphor in the sense that religion or world view is always a metaphor–an attempt to understand and maintain a connection with the larger life which, ironically, we can only understand partially–through a glass darkly. By all the measurements available to us, however, scientific investigation, meditation, and the evidence of sensory experience, the Gaia theory is the only concept sufficiently sound to use as the foundation for subsequent newly emerging precepts of ecological design.

Precept Two:
Design Should Follow, Not Oppose, the Laws of Life

Having designated the living Earth as the context for our thinking and for our ideas of design, it seems a logical next step to rejoin Charles Darwin in asking questions which we quoted in the last chapter. What, indeed, are the laws of life? Since our present concern is the design of future settlements, it is all the more relevant to examine Darwin's question in the context of what is, and what is necessary for life as we know it to continue. Just as the Gaian reality, the metapattern, exhibits characteristics of a living being, understanding the patterns and laws of the given living world, in as much as is presently possible, is fundamental. Within such a context biology is the model that mirrors most closely the workings of the natural world. Our second precept of design is therefore that *biology is the model for design*. With this in mind it seems appropriate to turn to the natural world with a biologist's eye and see what can be gleaned for our purposes from careful observation and analysis.

i. The cell is the basic unit and building block of life. As such it is an entity complete unto itself. This is most easily pictured by bringing to mind the kind of one-celled creature frequently perused in introductory biology

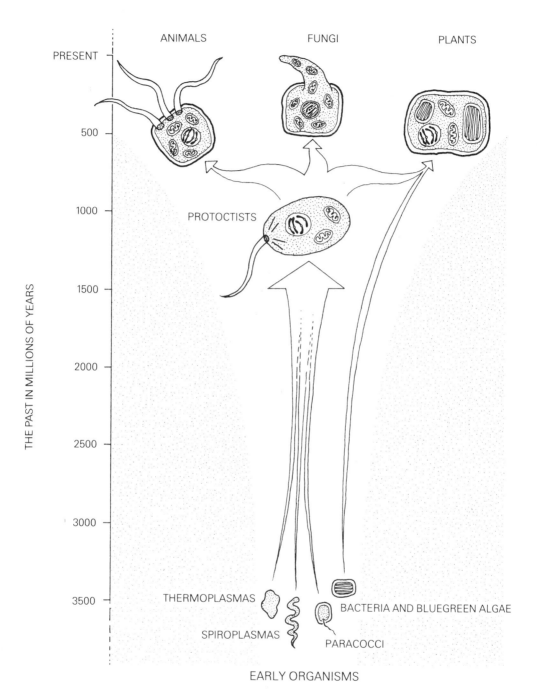

ANIMALS   FUNGI   PLANTS

THE PAST IN MILLIONS OF YEARS

PRESENT

500

1000   PROTOCTISTS

1500

2000

2500

3000

3500

THERMOPLASMAS

SPIROPLASMAS

PARACOCCI

BACTERIA AND BLUEGREEN ALGAE

EARLY ORGANISMS

Symbiotic Evolution in Nature

Life in a Small Pond

courses, the most popular usually being the amoeba. Like all other simple and primitive organisms, it is self contained. It carries out, at the microscopic level, all the basic attributes of life, including food gathering, feeding, excretion, respiration, purification, and reproduction. The same is as true of a cell in our own kidneys, a butterfly's wing, or a sumac leaf, as it is in the independently inclined amoeba.

ii. The cell participates directly in the fundamental functioning of the whole organism. There is a high degree of interaction and cooperation between cells. At the dawn of life on Earth, evolution was triggered when different kinds of organisms invaded or enfolded each other to produce composite organisms ranging from fungi and lichens to the higher plants and animals. Over eons of time the cooperation and interdependence between formerly unrelated organisms has grown ever more complex. Basic to biological functioning has been the bringing together of normally distinct organisms, the combined activities of which create new organisms. Concomitantly, many of the ancient, less complex precursors of present life forms, like bacteria, spirochetes, and blue green algae, still persist as independent entities. As such they continue to play a basic and major ecological role in the overall metabolism of the planet.

iii. The fact that organisms are at once complete, independent and autonomous, yet interdependent with other life forms, is a paradox to basic life. However whole and complete its structure, no organism is an island unto itself. Nature depends upon connections through different levels of biological organization. The connections are always immediate and near by. There is an unbroken continuum from cell to organism to the larger ecosystem and beyond to the bioregion and on again ultimately to the whole planet. Further, although, through differentiation, related cells become organisms that range from insects to trees, ancient biological patterns are not abandoned but maintained through vast reaches of time. In this way nature is extremely conservative—and this characterstic is a unity that permeates all of life.

iv. The ecosystem is the next level of organization and is analogous to an organism, the differences being that the boundaries are less distinct, the length between the components longer, and the couplings looser. An ecosystem is an interacting system of living organisms and their non-living environment. In a sense, the environment is the home within which organisms live. A pond is one of the simplest ecosystems to visualize because it is contained in a bowl of land and its boundaries are easily discerned. An ecosystem can also be defined in terms of contained relationships—the ecosystem of the food chain or the relationships of the essential gases which are controlled by organisms are examples. When a pond is exposed to sunlight, the algae give off the the oxygen essential to the survival of the animals. The bacteria and animals produce carbon dioxide which the algae and other plants need in order to live. Populations of smaller fish are kept in check by predators. Predators in turn are regulated by even larger predators like herons as well as by their own reproductive biology in that they produce fewer offspring. To make an anthropomorphic evaluation of such an arrangement as dog-eat-dog would miss the deeper meaning of nature. Just as all of us must live to eat, all life forms consume others yet also have a function beyond their own particular existence. In the end, all life is eaten or decays, that new life may be born and the larger life continue. Whereas organisms are outwardly defined by a particular structure or surface or architecture, such as bark or skin, or scales, topographically ecosystems are defined by the diminishing or outer limits of the relationships. While the definition of boundary in a pond is contained by the banks, more often one ecosystem, like a field, will blend into others like that of a wood or a neighboring lawn. An ecosystem is not just an assembly of creatures but, because of the integrity of its structures and the mixed relationships, it is a definable entity, a meta-organism. Just as relationships in animals are expressed through a central nervous system, an ecosystem like a pond expresses relationships as a gestalt—as the sum of its parts acting in dynamic concert.

v. Nature is not static. The natural world lives in flux and understands change. In a wooded area an abandoned lawn left to itself reverts to a meadow and then, within brief decades, to a woods. During this period, technically termed ecological succession, structural changes take place. The landscape becomes more diverse, stable and often less vulnerable to perturbations. In contrast more humanly derived systems, most of our towns and cities for example, indicate a frame of mind that could be called early successional. Structural relationships are defined and fixed at the outset and the pattern is hard to change as conditions change. We tend to build, destroy, rebuild, destroy and rebuild again. Too often we lock ourselves into inflexible designs which inhibit maturation in a given society or community.

iv. The bioregion, beyond the ecosystem, is the next over-riding structural unit, forming a cluster of ecosystems arranged topographically and climatically to produce a distinct region. A bioregion is easy to recognize but hard to define. It can be framed by a great river valley, by mountain

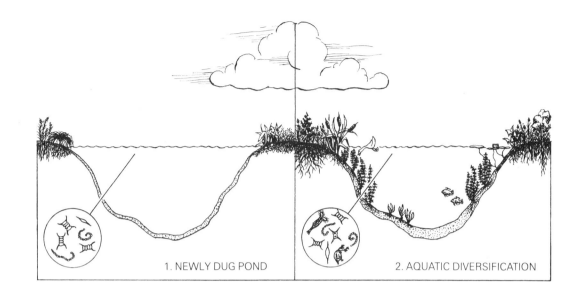

1. NEWLY DUG POND

2. AQUATIC DIVERSIFICATION

Succession in a Small Pond, Sequence Measured in Decades

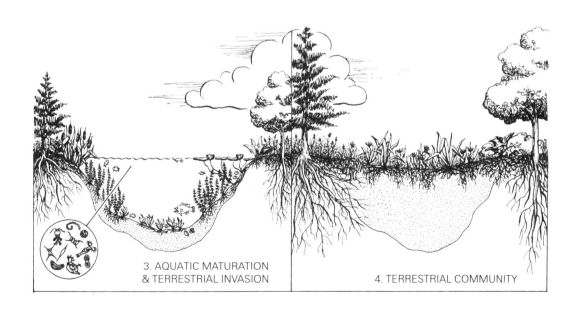

3. AQUATIC MATURATION
& TERRESTRIAL INVASION

4. TERRESTRIAL COMMUNITY

ranges or a coast. Usually it is categorized by distinctive vegetation and climate. Yet even a bioregion is not an island unto itself, for it blends outward to join with others to comprise a biographical province. The hardwood forested land east of the Great Plains extending southward from southern Canada almost to the Gulf of Mexico is one such great province. Such provinces in turn interconnect and blend to form the Earth's canopy—until eventually we again come around to Gaia.

vii. All life shares the same basic information. Humans, frogs, and mushrooms, all are built of the same matter laid down in slightly different combinations. Modern biology is revealing the language of the genes, the code which provides the informational or instructional framework for the unfolding creature. The informational attribute of nature brings home the realization that life cannot be analogous to a machine. Working with living material is completely different from working with lifeless subjects. Preparing a window box, designing a cluster of buildings, or reshaping space in a vacant lot can create unpredictable results. If a window box is innoculated with a few handfuls of forest soil, and contains flowers, herbs, and vegetables, if it is occasionally watered with water from a wild pond, it will unfold according to its own instructions. It will function as a magnet for unexpected forms of life and be delightful and informational as well as useful. There will be wildness in it. Something comparable happens when buildings, parks, and perhaps even towns are designed from ecological rather than purely technological blueprints. There is a qualitative difference that we can feel.

viii. Time, in nature, is more complex than time as we experience it. It is marked by the seasons which are linked to the time of year and contains life spans that may range from minutes for some microorganisms to centuries for certain trees. Beyond this is ecological time, called succession, which is usually measured in decades—although the range here can be enormous too. Beyond the ecological time is evolutionary time, usually measured in time frames ranging from centuries to millions of years, which describes the appearance, changes, and often the extinction, of life forms. Finally, we reach geological time, which overlaps with evolutionary time. This huge yardstick measures major physical events such as the formation of mountains and the drifting of continents, as well as the major climatic epochs.

We say that time is ecosystem-specific in regard to succession. Time marks the rhythm of the relationships within the ecosystem. Some microbial communities it contains may experience time in hours; the forests may measure successional time in centuries. As ecosystems unfold, succession leads to change, maturation, and increases in diversity and complexity. Succession is described in stages encompassing "birth", "rapid growth", a phase of structural richness, maturation, and finally, decay. Even forests grow old. The orderly progression toward the full expression of place is dependent on climatic, geological, and edaphic or soil factors, and on external forces such as fire and prior human presence.

Succession is a powerful conceptual tool for thinking about, designing, even reshaping communities. It allows us to cope creatively with change and even to steer it. In nature change is a creative force. We can use the same principles in design. In an ecosystem, succession incorporates an increase in diversity from a few simple organisms to a population of inhabitants highly evolved in complex associations. Such diversity increases the number and, more importantly, the kinds of relationships, thereby lessening links of rigid dependency by spreading them throughout a more complex community. Diversity leads to increased stability, protection from external change, variety, and overall system efficiency, which in turn results in greater order and information flow. The ordered complexity that is created embodies two attributes that scientists rarely discuss in their analyses, harmony and beauty. These are not ephemeral qualities. They have meaning that speaks directly to us.

At New Alchemy the imperative that biology should be the model for design was a given almost from the beginning. Aware, even in those just pre-OPEC days that we were a petroleum-addicted society, we were determined to design food producing systems based on renewable sources of energy. As we proceeded with our intent of growing food using ecologically based techniques, one fact soon became painfully apparent. On Cape Cod, which in the last century was largely cleared for farming, so much so that Thoreau on his tour of the Cape remarked somewhat acerbicly on its dearth of wood, the Gaian impulse is still undaunted in its determination to grow trees. Whenever human vigilance is relaxed in lawn or garden a pattern of succession is initiated that is clearly intended to reestablish a forest. Forests are the dominant landscape for the entire northeast, in fact, but ironically the inclination of agriculture, subsequent to that of the Native Americans, has been based on destruction, rather than the preservation or modification, of forests. In defining our longterm agricultural philosophy for New Alchemy, we decided that the biology of the forest would be our design model.

In analyzing the structure and functions of New England forests, we came up with an impressive list. Forests control erosion, moderate seasonal pulses in hydrological flows, buffer climatic extremes, and provide fuel, food, and wildlife habitat. To conceptualize a farm that reflected the image of the forest, the primary break with current agriculture was that the main source of energy would be the sun rather than petrochemicals. Our agricultural forests evolving with time would move naturally through succession, from cleared areas to forest. To remain most viable the farm could not be static but will mature gradually to a state comparable to that of a climax state in nature. Conceptually our farm begins at the bottom of the numerous fish ponds, and extends upward through the water to the ground cover formed by the vegetable and forage crop zone where livestock graze. It then rises through the shrub layer to the canopy formed by the trees that produce fruit, nuts, timber, and fodder crops. Following this plan we are hoping to maintain the farm in a dynamic state of ongoing productivity while it continues to evolve ecologically in the direction of a forest.

There are several instances of analagous research in other bioregions. In the Great Plains the prairie is the model for Wes and Dana Jackson at the Land Institute in Salinas, Kansas.[1] They are searching for and breeding perennial grains, including rye, sorghum, wheat, and corn that would fit into the ecological structure of a natural prairie. These new plants would take care of themselves as the prairie plants do. In the Sonoran Desert bioregion of the southwest, the oustanding work of Gary Nabhan of the Southwest Crop Conservancy Garden and Seed Bank is following a comparable course, in the image of the desert.[2] He is grafting his knowledge of Native American desert farming practices with modern ecological research to create a sustainable agriculture adapted to the pulses, climate, and soils of the desert. Although the continent separates us, the successional strategies we employ are the same as those used at the Land Institute and the Southwest Crop Conservancy. Only the exact agricultural form grows out of a regional ecological imperative.

When, in the evolution of New Alchemy, we became interested in proceeding from the agricultural landscape to the integration of agriculture and architecture, again biology was the model. We asked ourselves: What is the smallest contained living system that works completely on its own? Our only answer was the Earth. Inevitably, the life science of biology became the model for our first bioshelter. Looking at the question of how the Earth works we realized the need for an equivalent to the atmosphere to act as a solar collector. It was obvious that Buckminster Fuller had been working with atmospheric architecture in building his geodesic structures based on the Earth's great circle arches. We first chose, therefore, a geodesic structure with a thin transparent membrane or skin to serve as a collector of solar energy. The question then became: Where does the energy go? In the case of the Earth the answer is, obviously, into the oceans, without which the Earth would be totally unliveable mostly because wild oscillations in diurnal temperatures of desert area would be unmodified by the seas. We calculated that, as seventy percent of the Earth's surface is water, we would approximate that relationship in the bioshelter. We dug a subsurface pond to act as a buffer for swings in temperature.

The next item on our agenda became a bit more complex, for we were interested in having our small body of water mimic the workings of the oceans. The ocean remains healthy and viable because of upswellings, like the Peruvian Current and the George's Banks, which rise the surface laden with nutrients from the colder lower depths. To simulate the workings of the ocean we used components which would also serve the dual function of providing us with a food crop. We introduced mirror carp, fishes which have such a powerful swimming motion that they stir and overturn the water as they swim—which causes the lower water to rise to the surface and come into contact with the warmth from the sun. The oceans have techniques for filtering as well as mixing—great blue whales are among the most impressive examples. Our equivalent was to introduce into our pond a miniature version, the tilapia or St. Peter's fish. The basis of the tilapia's diet is the algae which we had added to the pond water, because algae are the major source of gases for the Earth's atmosphere. The tilapia swam around the pond with their mouths open, like small whales, filtering the water and fattening for harvest. We brought in yet another species of fish to substitute for the function of the rivers in the role of bearing terrestrial nutrients to the oceans. We used the white amur, which feeds on vegetative matter like grass clippings and flower stalks and contributes these ingredients to the pond by passing them through its digestive tract. Collectively, our polyculture of fish made the pond independent of the fossil fuel-driven equipment which otherwise would have been necessary to serve the purposes of mixing and filtering the water, as trapping the sun had freed the structure from the need of fossil fuels for heating. We maintained a continuity with the Earth's seventy/thirty percent ratio of water to land in this early dome. In the border of land around the pond, we planted a garden which fluctuated in its crops with the changes in season. Although many of the components of that early dome have been modified in later bioshelters, the fundamental procedures of looking to the living world for inspiration and of using biology as the model for design, has remained constant.

Precept Three
Biological Equity Must Determine Design

Saul Mendlovitz of the World Policy Institute has a question he is wont to ask when taking part in meetings or conferences that are engaged in examining trends or possibilities for the future. Whether the issues under discussion are political, economic, or technological, he asks: How will it affect the poorest third of humanity? However obvious and urgent this question may sound, from the worsening plight of the poor people of the world, it is evidently not posed or seriously examined frequently enough when a new policy or venture is unleashed by the developed world upon them.

Mr. Mendlovitz has not permitted us to disregard his concern. He questioned us the first time he heard of our work at New Alchemy and has caused us to keep the future of the poor of both developed and under-developed nations constantly before us in our design work. Saul Mendlovitz's views of our work was reinforced when Margaret Mead made her comment that our design with bioshelters had, at the time of her observation, little relevance for poor people. The longer we worked the more it became self evident that biological design and biotechnology could not be divorced from issues of social justice. Biological equity, the just access to and distribution of basic resources is, unavoidably, a precept of biological design.

The abuses of this precept are far more numerous than the instance of it being honored. Francis Lappé and Joseph Collins in *Food First: Beyond the Myth of Scarcity* have documented how the agricultural self-sufficiency and nutritional basis of Third World peoples is undermined through the infiltration of multinational corporations interested in expanding their markets or exploiting their resources to create luxurious and exotic products for sale to the developed world. The Worldwatch Institute has published more than fifty books on the depletion of basic resources, particularly in the Third World, which has led to deforestation and desertification, and the increasing unavailability of fuel. The urgency of these problems for the people affected is matched only by their complexity.

An awareness of world hunger and malnutrition were very much with us when we put together our first aquaculture programs at New Alchemy. J. Baldwin, who is one of the pioneers of appropriate technology and was working with us at the time, was inclined to look at the question of social justice from another angle. He used to remind people at the Institute that, at the end of every wrench not to mention far more imposing forms of technology, were the steel mills of Gary, Indiana: implying that social and technological issues are as old as our use of tools and are probably forever fatefully entwined. With one of our most recent projects we have made a broader attempt to tackle the knotty problems of social and biological equity. Rather than keeping a concern for Third World peoples at bay or living with it as a nagging worry, we decided to try to work closely with a Third World country and, putting our accumulated expertise and knowledge of biotechnology at their disposal, to undertake a project that had energy, technological, economic, and nutritional repercussions.

From about 1976 on New Alchemy's work began to be sufficiently recognized for many of us to be invited to travel internationally to discuss the implications of the work. Working in coastal areas and on islands we had a chance to observe yet another horn of the dilemma that industrialization and an era of cheap fossil fuels had inflicted on coastal and fishing peoples. Visiting artisanal fisheries in the Indian Ocean, in the South Pacific, and in the Caribbean, we found that rising fuel costs, supply disruptions, and the unavailability of spare engine or fishing gear parts, not to mention the disappearance, through rapid deforestation, of traditional boat building woods, had idled many offshore fleets. We learned that in many places traditional crafts had been abandoned and modern fishing boats adopted in the period between the end of the Second World War and the early 1970s, which had been a time of worldwide economic expansion characterized by low interest, easy credit, and expanding world markets. With the subsequent global recession and the accompanying diminution in foreign credit and foreign exchange resources, the then-expanded fisheries became vulnerable to the vagaries of international fuel and finance sources over which they had virtually no control. The need to rethink the basis of transport for coastal peoples was becoming increasingly obvious.

The impressions that we had been gathering from our own travelling were reinforced by reports from other parts of the world from a number of our colleagues. Before her death Dr. Mead was drawn to the idea of sail-powered shipping which, as the tall ships have proved, have a power of imagery that is difficult to explain. The heart is moved, and perhaps the soul

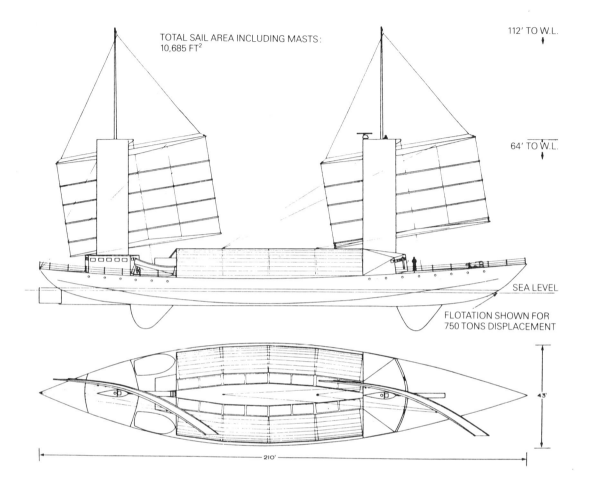

TOTAL SAIL AREA INCLUDING MASTS:
10,685 FT²

112' TO W.L.

64' TO W.L.

SEA LEVEL

FLOTATION SHOWN FOR
750 TONS DISPLACEMENT

43'

210'

Ocean Ark, the "Margaret Mead"

as well, by their elegance and by the incipient sense of adventure that stirs at the sight of swift moving sails on the horizon. More than anything else, they seem to symbolize the possibilities of rising above rather than being trapped by the exigencies of the post-petroleum era.

We spent quite a long time mulling over ideas of biological restoration, the potential of biotechnology, and the return of commercial sail. In 1981 we founded Ocean Arks International as the vehicle for implementing these concepts. We had thought at first about the possibility of a sort of sail-powered greenhouse, a "Biological Hope Ship." The idea was that the boat would produce and transport biological materials like seeds, plants, trees,

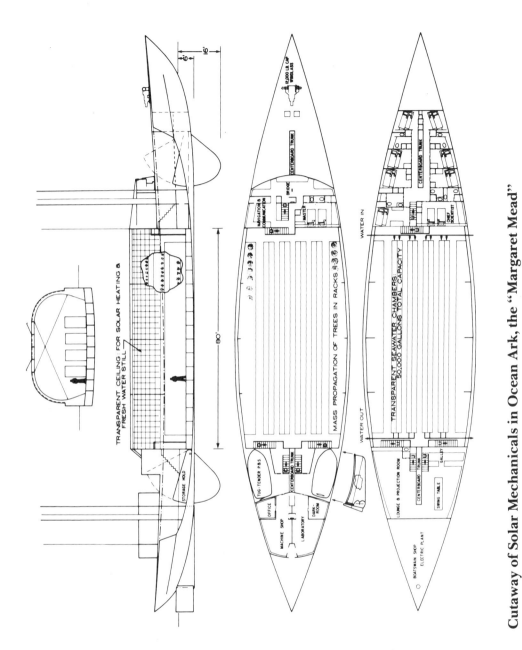

Cutaway of Solar Mechanicals in Ocean Ark, the "Margaret Mead"

and fish to impoverished areas with the hope of reviving the local biological support base and thereby improving the means for the human population to sustain itself. The naval architect Philip Bolger designed the 210' long Ocean Ark, the "Margaret Mead" to our specifications, and we built a 50' long ¼ scale vessel to test some of his modern rig concepts. After considerable experimentation and thought, however, the costs and the technical innovations predicated by a project on so large a scale led us not to abandon it, but to tack off in a different direction. We decided that it was probably more feasible to start conceptually with smaller vessels and work up. On a visit to Marlon Brando's island off Tahiti, we had been impressed that Marlon's son, on a Hobie Cat, could far outfish the rest of the villagers in outboard motor boats. It seemed to indicate that the future lay in the speed of sail rather than the availability of diesel fuel.

We knew of the naval architect Dick Newick by reputation as the designer of some of the fastest racing yachts in the world. His trimaran racers hold many transoceanic records and in 1980 "MOXIE", his fifty-five foot trimaran, sailed singlehandedly by Phil Weld, who was then sixty-five, won the TransAtlantic OSTAR Race. We learned that Dick Newick had also pioneered a number of new nautical technologies and construction methods. That he might be sympathetic to our kind of thinking was indicated by the fact that he had also designed "SIB", a lightweight but sturdy sailing fishing craft named after E.F. Schumacher's *Small is Beautiful*.

After several meetings and discussions he came up with the right design for the kind of boat we had in mind. Together we agreed upon the plan for what we decided to call the Ocean Pickup, intending it to have the same kind of wide ranging usefulness as its land counterpart. We understood that to have any appreciable impact, the Ocean Pickup could not in any way seem like second-hand technology. It would have to be as fast or faster than a motor-powered boat and just as advanced. Auspiciously, over the last decade there have, in fact, been four related technological developments which, taken together, represent a major breakthrough that could make such a boat, as well as the rebuilding of artisanal fishing fleets, possible. In the main the building material for such boats is intended to be ubiquitous, non-traditional, fast-growing scrub trees.

The first of the innovative technological developments employed by the Ocean Pickup is a wood/epoxy saturation technique or West System developed by the Gougeon Brothers of Bay City, Michigan, which makes it possible to use softer, light woods than have been normally used in boat building. With this technique strips of wood are used as a structural fiber which, when layered together with epoxy resins, forms a composite engineering material that is rot-resistant, light-weight, has a high strength-to-density ratio, and is very resistant to fatigue. Dick Newick has tested one South American softwood, Baromalli (*Catostemma commune*), from Guyana, by making sample boat panels which proved exremely well suited to wood/epoxy boat construction. The epoxies are not expensive and would come to about ten percent of the cost of a fishing boat in most Third World countries.

The second of the new techniques emerged from an attempt to get around the problem that wooden boats are traditionally custom-built and therefore expensive. Dick Newick came up with the idea of a master mold that would permit "families" of shaped panels to be easily fabricated. Different panels from the same mold can be used for a variety of boats, for both the decks and hulls. James Brown, an associate of Dick Newick's, uncovered the geometry necessary to permit compound curved surfaces to be covered with veneer strips so that each of the strips has an identical profile shape. The technique is called Constant Camber. It enables the fast and inexpensive fabrication of compound, curved wood panels. Boat hulls can be made up of mirror image cold-molded panels which can be mass produced, and will therefore be competitive economically with steel, aluminum, and fiberglass. The third of the new developments used on the Ocean Pickup was vacuum bagging. Vacuum bagging is a technology that allows the marriage of epoxies and wood over large, compound, curved surfaces so that veneers can be laminated into hull-shaped panels. Vacuum bagging evacuates the air from between the layers of veneer and permits the thorough penetration of the epoxy. The result is a composite boat-building material that is strong, ultra-light, rot-resistant, and conservative in the use of epoxy, which is why epoxy represents such a small percentage of the overall cost of a boat.

As originally developed, cold-molding vacuum technologies were capital intensive and industrial. Then an engineer named John Marples collaborated with Dick Newick and Jim Brown to convert it to a low-cost "backyard" technology. Vacuum bagging now can be done by villages of small communities throughout the world. The vacuum pumps can be built cheaply from old regrigerator units run backwards to create a vacuum. Yet another advantage to vacuum bagging is that there is no need for nails, staples, or screws in the hull or decks. As a result, weight and costs of the vessel overall are less, as is the dependency on imported fastenings.

The fourth and last of the innovative elements in the Ocean Pickup is also the most visible: the fact that it is a trimaran—a three-hulled sailing vessel. As a leading designer of multi-hulls throughout his career, Dick Newick has synthesized the most advanced aerodynamic and hydro-dynamic concepts with the most modern materials. With the Pickup, he has then simplified them so that they can be readily adaptable to working water craft throughout the world. Interestingly, Dick sees himself as working in a millennia-old tradition of boat design with its roots in the watercraft of the South Pacific and South East Asia. Now his Ocean Pickup has opened up another dimension for multihulls in the form of carrying capacity. It is designed to carry a ton and a half of cargo, plus crew, even though the overall weight of the boat itself is less than a ton. It is intended to sail at working speeds of up to twelve knots, fully loaded. With a light load it is even faster, sixteen knots or more. The Pickup does carry an auxiliary outboard motor for periods of calm but the fuel consumption of the motor is as low as two pints per hour at six knots. Thoroughly modern in design and materials, it is not intended for a career as a racing or leisure craft but as a workboat—a sea-going Pickup.

Extrapolating for a moment from a single Ocean Pickup to a fleet of comparable vessels, the economic prospects for a coastal fishery broaden enormously. Fishing and coastal people could find themselves in possession of a fleet that is largely self-reliant in terms of fuel and equipment. They would be able to re-establish control of most of the building resources, the technology, and the actual construction of their boats, as ideally, the boats will be built locally of fast growing indigenous woods relying minimally on imported items like the epoxy. The construction methods should ensure that the boats will be long-lived and require very little in the way of maintenance. The boats would be easily replaced when necessary.

There have been a number of expressions of interest in the Pickup from such countries as India, Indonesia, Sri Lanka, and China. From the manager of a fishermen's cooperative in Costa Rica, La Cooperativa de Pescadores del Liroral Atlantico, we received the following letter.

> Dear Sir:
> We are very much interested in doing sea trial of your one ton ocean pick up vessel.
> Our association is composed of 208 inshore artisanal fishermen (from the coast of Atlantic Limon, Costa Rica) with a lot of economical problems.
> We think this boat will be great help to resolve some of our critical conditions, such as high costs of fuel and replacement parts for use of our outboard. Owing to these facts, our activity became non profitable.
> We sincerely hope that you will give us the opportunity of testing one of these boats.

Also from Costa Rica, from Bill McLarney and his colleagues there, we have had further assurances of interest in a boat such as the Pickup. For some time NAISA has been experimenting with fast-growing trees. Geronimo Matute, one of the leaders of the Gandoca community, has melina (*Gmelina arborea*) trees on his land which after only three years have reached nearly a foot in diamter. Should the melina not prove the best tree for the wood/epoxy technique, Bill McLarney lists Albiizia, Sesbania, Eucalyptus, and several other fast growing local trees as possible candidates. The intent is that potential for reforestation will grow with a specific motivation.

It may be apparent that hidden in our agenda of bringing back working sailing craft for functional transport and transportation is an attempt at land restoration as well—just as during our early work we concentrated on the land, knowing that one day we would turn to the sea. Embodied in the plan for using scrub trees as a building material is the hope that the impulse to plant fast-growing trees which would be ready for use within a foreseeable time-frame, perhaps three to seven years, is more compelling than planting trees that take generations to grow. For most people actions with consequences beyond the pressing demands of the moment constitute an almost gratuitous act. In such circumstances, for land restoration or an ethic of stewardship to have any meaning, the results have to be perceived as achievable within a realizable period of time. Planting even weed or scrub trees in a world of deserts on the march is a workable first step. Even more quixotically hopeful, perhaps a few of the remaining giants of the forest may be spared in favor of smaller trees with shorter generation spans.

The first Ocean Pickup was built by Dick Newick in his shop on Martha's Vineyard during the fall of 1982. Like the openings of the Arks several years before, the launching in late November of that year was something of an occasion. We had invited the Very Reverend James Parks Morton of The Cathedral Church of St. John The Divine in New York City to bless the boat before it was launched. Noon was the time we had chosen because we wished to begin with United Nation's Prayer for Peace, a prayer intended to be said at noon every day around the world so that it is continuously circling the Earth. This helped to set the context for the launching of a boat which, like the bioshelter, is intended to help create an infrastructure of non-exploitative technologies to foster a peaceful world. The boat was named the "EDITH MUMA" for the courageous lady who first saw the relevance of New Alchemy's Ark, and with her husband, the late John Muma, funded it. Dean Morton blessed the boat, using biblical references to the great immeasurable sea and concluded, "Bless, we beseech you, this modern ark that it may be a showing forth of the abiding and friendly power of crea-

The "Edith Muma"

tion when used in peace, that wind and sail may again join together for the welfare of thy servants and the revelation of thy loving care."

It was very quiet for a moment as we listened to Dean Morton, many of us half praying, half willing, that the "EDITH MUMA" fulfill its intended promise and help to move us a little closer to peace with the living world and among ourselves. Then Margaret Lloyd, another close friend and affiliate of New Alchemy, opened rather than broke a bottle of champagne to christen the "EDITH MUMA". Not to break the bottle seemed a bit off and anticlimatic at first but it was the idea of Dick Newick and the crew, who thought

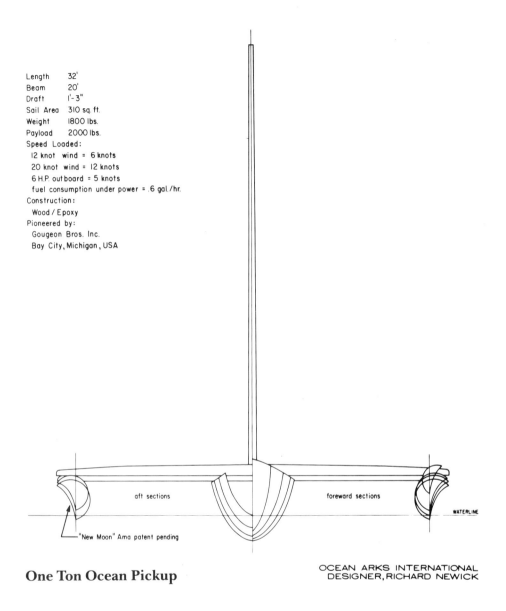

Length   32'
Beam   20'
Draft   1'-3"
Sail Area   310 sq. ft.
Weight   1800 lbs.
Payload   2000 lbs.
Speed Loaded:
  12 knot wind = 6 knots
  20 knot wind = 12 knots
  6 H.P. outboard = 5 knots
  fuel consumption under power = .6 gal./hr.
Construction:
  Wood / Epoxy
Pioneered by:
  Gougeon Bros. Inc.
  Bay City, Michigan, USA

aft sections   foreward sections

WATERLINE

"New Moon" Ama patent pending

One Ton Ocean Pickup

OCEAN ARKS INTERNATIONAL
DESIGNER, RICHARD NEWICK

a non-violent gesture that also would not litter the beach with broken glass better fitted the spirit of the day. They were right. Once Margaret Lloyd had splashed the boat with champagne, the entire crowd moved together to carry the Pickup over the sand and set her in the sea. The crew scrambled aboard and adjusted lines and rigging and raised the sails while Paul Winter, whose music has so often cried out for the preservation of sea creatures, under-

.384

.622

.238

By-Catch
Georgetown

2 TON OCEAN PICKUP

WATERLINE

OCEAN ARKS INTERNATIONAL
DESIGNER, RICHARD NEWICK

Guyana By-Catch Project, Two Ton Ocean Pickup

scored their movement with the music of his saxophone, extending non-verbally the blessing of Dean Morton.

After a season of testing for both sailing and fishing ability in the offshore waters of Cape Cod, the first Ocean Pickup sailed to Guyana in South America, a distance of approximately 3,700 miles, where, under the auspices of the Guyanese Department of Fisheries, she was tested by local fishermen. The "EDITH MUMA's" mission in Guyana was two-fold. Not only was she on trial with the fishermen for suitability as a replacement for the motor-driven drift-gill net vessels they had been using, she also was being

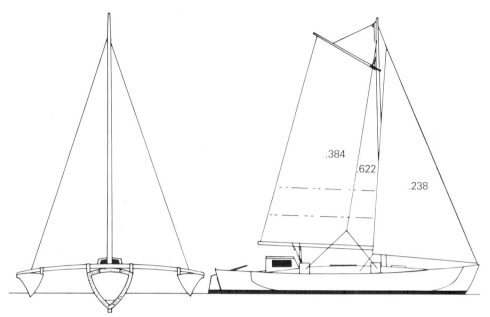

Rigs of Three Ton Ocean Pickup, Guyana By-Catch Project

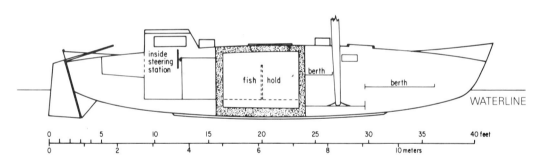

Guyana By-Catch Project, Two Ton Ocean Pickup, Showing Fish Hold

considered as a relay vessel. By far the most profitable catch for Guyanese fishermen is shrimp which they bring back to port for export to the developed world. When they spread their nets for shrimp, however, they haul back in a large number of other fish as well, often up to fifty percent of the catch. The fish caught this way are referred to as a "by-catch." The shrimp boats have a limited freezer capacity so the fishermen cannot afford to take up space on board with moderately priced fish as their economics are predicated on the expensive shrimp. The fish are thrown back into the sea, dead.

Although the Guyanese people do not go hungry, their nutrition would be greatly improved if this by-catch was not completely wasted. It is not cost effective to use trawlers or other motorized vessels to rendezvous with the shrimpers, so Fisheries officials decided to test the Ocean Pickup as what is called in local terminology a "by-catch buy-back boat." She has sailed successfully each morning out to meet the trawlers and load the fish from them into the deep hold, then raced back to Georgetown to sell the fish at reasonable prices. In this way it became feasible in 1983 to prevent the ongoing loss of an important food resource for the people of Guyana and the Caribbean basin.

Obviously, fleets of sea-going Pickups can never ease all the problems of Guyana or any other coastal country, but as a result of one Pickup a few more fishermen are going to sea and a few Guyanese have more protein in their diets. If enough future design took the welfare of if not the poorest third of humanity but only a small fraction of that, as integral from the outset, slowly the fate of the poorest third might become less desperate.

Precept Four
Design Must Reflect Bioregionality

In the last chapter we defined bioregion as a cluster of ecosystems arranged topographically and climatically so as to delineate a distinct region. At that point we were concerned exclusively with biological concepts. The human element was not under discussion. Obviously without a consideration of the human role, however, any definition is incomplete as, directly or indirectly, there is very little of the planet where we have not left in some way our mark. Nor is bioregionality an easy concept to define. It is a notion that has grown from uneasiness at environmental destruction, increasing cultural homogeneity, and exasperation at the sluggish ineffectuality of centralized governments and their attendant bureaucracies to address environmental issues and the complexity of modern life. Advocates of bioregionalism hold that it is a crucial conceptual tool for protecting and preserving the biological, cultural, and political integrity of a given area.

Writing in *CoEvolution Quarterly* for the winter of 1981, Jim Dodge stated, "Bioregionalism is simply biological realism...holding that the health of natural systems is directly connected to our own physical/psychic health as individuals and as a species. We must be constantly interconnecting our own worlds with other natural systems. Even so," he adds, "No matter how great our laws, technologies, or armies, we can't make the sun rise every morning nor the rain dance on the goldenback ferns."[1]

Dodge enumerates a partial list of criteria that determines what it is that constitutes a bioregion. A bioregion includes biotic shift, meaning the percentage change in plant and animal species composition from one region to another; watershed or system of river damage; land form or topography; the culture, perception, and behavior of the human population; elevation or altitude, and the force or spirit or over-riding essence of the place itself.

For most of humanity's evolution bioregionalism has been unself-consciously and effortlessly a part of design—from the yurts of Central Asia to the magnificent Pueblo dwellings of the American southwest, to the tents of wandering bands of nomads—culture and identity, geography, topography, climate, and indigenous resource base all have been for millennia silently but eloquently expressed in a manner appropriate to the bioregion. The contrast is great between the diversity of such structures with the recent trend toward homogeneity in cities world wide which have been erecting skyscrapers that are ringed with bands of urban sprawl. It is only because most of us do not have a sense of human history over this long period of time that we do not feel how *odd* it is to build cities and suburbs as we do.

Taking a very broad and much simplified overview of the North American continent is one way to illustrate a number and variety of bioregions. It is at once obvious that each of them portray, in the broadest terms, the meaning of climate. The forests of the northeast indicate the predominance of moisture there. In New England, the diffuse light from the sky is buffered by the North Atlantic Ocean and is filtered and muted. Toward the center of the continent are the prairies, natural grasslands which are similar to a terrestrial sea in their vastness of movement and light. Distance is omnipresent, filled and rimmed by sky and winds with sun and dark intermittent rain clouds building far off. The deserts of the southwest convey intensity—searing heat, glaring sun, deep night cold, and raw architypical forms—sculpted beauty beyond the human hand. The northwest is again mainly forested and damp but different from the east coast in biological make-up and over-riding essence. Farther south, the diversity of California's climate and topography is reflected in the diversity of its cultures.

In the northeast, before the coming of Europeans, Native Americans lived in quonset-like, long houses made of saplings and layers of bark. The arhitecture fit—was suited to the climate. A recent computer study of the optimal shape for a low energy, advanced solar bioshelter was uncannily reminiscent of such quonset-type long houses. The model indicated that the structure sited in an east-west direction be insulated and reflective on the north side and north side interior respectively, and transparent to light on the south side. Except for the transparent glazing, this type of building, following our design precepts, could have existed centuries ago.

Early European settlers in New England had ideas other than those of the Native Americans. Their houses favored high-pitched roofs, salt box shapes, and many fireplaces. They were fairly energy-conserving structures, but the open fireplaces burned copious amounts of the then plentiful wood. Surrounded by forest, the settlers were wasteful and as a result the original forest had disappeared within two hundred years. By the nineteenth century houses were no longer sited according to the dictates of climate but faced the street or road and reflected wealth and status. Some of these were elegant with an almost timeless purity of line, but they still required considerable heating fuel. Increasingly, the principal fuel was coal, which at the relatively low temperatures of firing in household furnaces or stoves, was a polluting material. By the middle of the twentieth century climatic sensibility had been utterly abandoned. Made possible by cheap fuel, the inexpensive, sprawling, badly-sited, fuel-consuming California ranch house had moved east and north. In a colder climate, the extensive use of glass, spread-out configuration, and single story architecture demanded extra heating and cooling. Picture windows which faced north as often as not, further exposed such houses to heat loss and chilling winds.

Before the westward march of civilization the teepee was the dominant traditional architecture for the plains. In a region where hunting grounds shifted and firewood was hard to come by, the teepee was light, portable, energy efficient, and seasonally adjustable, not unlike soft, architectural extensions of clothing. As farmers, the early white settlers were stationary. Livestock and grains necessitated storage, so they built sod houses, barns, and outbuildings. Some of them were well bermed, with sod and dirt, and faced south for light and supplemental heating. Of necessity, as wood was scarce, they were low-energy houses. But by the middle of the twentieth century, prairie architecture caught up with the rest of the country and here again, the California ranch-type house came into vogue. Prairie architecture, however, could take another course. If traditional notions were grafted to ecological ideas, an indigenous housing which grows out of and is part of land forms could be developed. These could possibly be in the form of a marriage of bermed and sod-covered structures looking like hillocks, to light solar forms rising out of the earth in teepee shapes.

The extreme climate of the deserts has produced some of the most beautiful and powerful clustered architecture in North America. For over a thousand years Native American builders developed and refined their designs with nature. Hopi builders were masters at minimizing the effects of climatic extremes. Their settlements were built into massive, south-facing, rock faces that trapped radiant energy when the winter sun was low in the sky. They used thick, heat-absorbing adobe as a building material to buffer the extremes of heat and cold. After the arrival of the Spanish, villages and pueblos became less well sited partially because, in some cases, military considerations were predominant. The natives and settlers did continue to use adobe clay as a building material, recognizing its function as a thermal buffer in preventing overheating in summer and extreme chilling in the winter. Thick walls, proper solar orientation, and suitable lighting persisted in regional architecture and some of this tradition yet survives. But again, by the middle of the twentieth century, the most prominent form of urban architecture had become the ranch house.

The ranch house in its original form and indigenous surroundings was an elegant and appropriate architectural form, its design reflecting bioregional cultural, climatic, and resource criteria. As long as we lived in an era of cheap fossil fuels even its cloned descendant offered the advantages of convenience and relative economy. But we now no longer live in an age of innocence with regard to energy, and the inadequaces of inappropriate design are becoming increasingly apparent. What has become of primary importance now is how we think about design. Merely to substitute an ameliorated form of the ranch house or the high rise is to limit our thinking to the short term and is a continuation of the same process that has us in trouble now. In mathematical terms it would be a change of coefficients but not of structure.

The concepts of bioregionalism, in concert with the other precepts of design we have described, offers a means of learning to think in a different and more integrated and comprehensive way. In launching into a description of bioregionalism in relation to design, however, we are in danger of finding ourselves hoist with our own petard. We cannot possibly cover the bioregion or even its facsimile, of every person we hope to reach as a reader of this book. That, eventually, must come from the inhabitants of each area—not from outsiders. We shall, rather, heed William Blake's caution that, "General good is the plea of the scoundrel, hypocrit and flatterer," and confine ourselves to what Blake called "minute particulars."[2] We will limit our discussion of the role of bioregionalism in design to our own bioregion of Cape Cod and show its functioning by examples, only venturing briefly afield to describe a design appropriate to an area in Colorado that we have come to know well.

We live on Cape Cod, a former peninsula that juts into the mid-Atlantic, made an island early in the century by the exertions of the Army Corps of Engineers when they dug the Cape Cod Canal. The sea is the over-riding natural presence, an ameliorator of climate and the only untamed wildness. The sandy, acid soil, the remains of a terminal glacial moraine and our legacy from the ice age, is covered with pitch pine and scrub oak and dotted with kettle hole ponds. Stone walls straggle through the low-growing woods, memorials to the sheep farms of the last century. Long a home of the Wampanoags, for many years after European settlement the Cape was an agricultural area supplying Boston with vegetables and fruit, even growing some grain. Now, with the exception of fish and native cranberries, strawberries, and blueberries, we import almost all our food at extremely high prices. Cape Cod has become a slightly rustic, woodsy suburbia, and a haven for tourists. The poverty of close to a third of the inhabitants is fairly well disguised in the general affluence. It is an area ripe for increased independence in food and energy.

According to the *Barnstable Register*, one of Cape Cod's leading newspapers, the past decade has done more to alter the character of the Cape than the previous two hundred years. The newspaper has kept a close eye on the Cape, describing the pace of development here as "virtually unmatched in America." Arthur Palmer, author of the book *Toward Eden*, and an environmental lawyer and landscape designer, has pronounced Cape Cod "a community on the edge of a cliff"[3] because, again in his words, it is an "attractive nuisance." Its attractiveness has led, as a result of very rapid population growth, to haphazard commercial development, traffic congestion, problems with waste disposal, and water pollution. This last, along with the threat to the remaining agricultural land, present together the most pressing ecological problems. Whereas most bioregions think in terms of watershed, on Cape Cod our water source is a single aquifer which lies like a lens under the ground. It can only be replenished by rainfall. The growing population is drawing off water more rapidly than the rain can keep it filled, so the level of the water table is slowly going down and, in the foreseeable future, could be inadequate to meet the needs of the Cape's inhabitants. Furthermore, although there is little industry, as a result of run-off from years of spraying with pesticides, fertilizers from the cranberry bogs, salting the roads in winter, leaching from septic tanks, and some leaching from the dumping of toxic wastes by the military, the purity of the water is in question and certainly is seriously threatened if nothing is done fairly soon.

The soundest solution would seem to lie in bioregional planning. In one sense, such a concept is not too difficult to communicate on the Cape because, since the cutting of the canal, we are an island and our boundaries are easily recognized and defined. With characteristic Yankee independence, however, the various committees and planning boards of the Cape's fifteen towns have always met separately to consider, separately, such common problems as water quality, land use, and the despoilation of the natural environment. Recently with considerable encouragement from several newspaper editors and community leaders, Cape-wide meetings and workshops have been organized and have succeeded in achieving a fair hearing for bioregional ideas. The word "bioregion" is now tossed about quite casually in the local press. Greg Watson of New Alchemy has proposed "A Cape Cod Regional Development Plan," the first step of which would be to develop a complete inventory of local natural resources. He has made a proposal reminiscent of New Alchemy's early work in the California hills, through which he intends to catalogue and codify information on the soils, geology, hydrology, and vegetation on the Cape. It is Mr. Watson's hope that such an inventory, when complete, will influence zoning laws, subdivision bylaws and building codes. Accepting the fact that development here cannot be stopped, the inventory is intended as an educational tool through which it may be limited and controlled.

Whereas this aspect of New Alchemy's recent work has been more political in terms of bioregionally appropriate design, there are now also several advanced new structures at the farm which can serve as models of design suited to the bioregion: a super-insulated auditorium designed not to need central heating and a pillow dome which will be described later. New Alchemist Ron Zweig suggested to the Town of Falmouth that it has the potential to improve its water purification capability by establishing a treatment plant based on a bioshelter design which would provide a solar heated environment for sewage-purifying aquatic plants. A full-scale ecological solar powered waste treatment plan along this design is currently under construction in Woodstock, New York, for which New Alchemy was a consultant. Falmouth will not likely follow suit as we did not come forward with our ideas until another commitment was too far advanced to change plans, although town officials did express some regret. In the long run our solar powered design would have been less expensive in terms of energy costs and would have offered the further economic advantage of commercial by-products in the form of soil amendments and livestock feed made from processing the aquatic plants and biogas for use as a fuel.

Although integration with its own immediate community and surrounding bioregion has been a part of New Alchemy's plans since the beginning, acceptance on the home front has been the hardest to win. In the winter of 1977, however, came both the opening and the organization to allow us to serve our own community more directly. We were approached by the Community Action Committee of Hyannis, a group with a long and established reputation, to band together with four other groups, including the Housing Assistance Corporation, the Wampanoag Councils of Mashpee and Gay Head, and the Martha's Vineyard Energy Resource Group, Inc. The result of this coalition was the Cape and Islands Self-Reliance Corporation. It has a claim to uniqueness in being the first of a kind—the first food and energy assistance corporation dedicated to low income residents of a particular area in the country.

The Co-op, as it is familiarly called, has given New Alchemy a timely opportunity to share the knowledge and skills that we have been accumulating over the years, and equally important, a network for reaching people, often those who most could use it, who otherwise might not have known of our work. The program which has grown out of the collaborative effort offers free energy and food audit and financial counseling for home energy improvements and food production. It provides access to the skills, services, and materials needed for energy conservation, renewable energy use, and local food production at discount rates. It also holds technical workshops with courses that cover the ecological, political, social, economic, and legal considerations integral to the effective functioning of small local organizations so as to further empower them to greater independence.

Largely as an exercise in ecological fantasy we have created a design for an area much more modest than that of the bioregion of all Cape Cod. It is for Sippewissett, the area that immediately surrounds our house. Sippewissett is a part of the Town of Falmouth. Around the turn of the century it was a combination of sheep farms and a summer colony. Now almost each one of its since wooded acres has a house on it. Rimmed with superb beaches the area attracts tourists, and the fine salt marshes are a magnet for marine biologists. There is one sprawling Victorian hotel which serves group tours in the summer. Four-hundred acres of woods have been set aside as a preserve to provide wildlife habitats, a haven for townspeople, and a place for outings and for running in the spring and fall and for cross-country skiing in winter. Sippewissetts's inshore waters provide some sport fishing and good clamming; a few lobstermen work the offshore rock formations. It has one riding and horse boarding stable and an excellent little orchard and market garden farm which does a booming business. There is some boat building and repair work carried out by a few master craftsmen in their backyards. Commercial zoning begins a mile or two away closer to the center of Falmouth. The most likely course for Sippewissett lies in maintaining the status quo. At the present, during the summer at least, the area is close to its human carrying capacity. A more promising and stable future calls for the preservation and enhancement of the woods, the salt marshes, and inshore marine life.

This could be achieved by the residents undertaking to diversify the employment base of the area. Our idea is to turn to the sea and create floating farms offshore from the great Sippewissett marsh where vast numbers of oysters, clams, mussels, striped bass, salt water-acclimated trout, and salmon could be cultured. Most of the technology to accomplish this has been developed in Japan and Scandinavia, and some elements have been tried successfully in North America. The Cape, with its great range of currents and sea temperatures could be a major center for the culture of marine foods. Objections to floating rafts marring the view of Buzzards Bay could be ameliorated by towing the complex offshore in summer. The food culture rafts could be seen as an alternative to further housing development and resultant pressure on the immediate ecosystem and more broadly, on the Cape's questionable water table. A program to produce trees useful in boat building, both fast growing trees for wood/epoxy boats like the Ocean Pick-up and more traditional building trees, would pay within a generation and would further diversify and preserve the woods.

Realizing a design such as this for Sippewissett is still unlikely, but some elements already exist. It helps that we are very sympathetically treated by the local press, particularly the *Falmouth Enterprise*, which has been extremely supportive of New Alchemy. One story we did about the possibilities for Cape Cod in the year 2000, which was predictably an ecological scenario, was read widely and commented upon. Already germs of the ideas are floating about in the air like dandelion seeds. Some have already taken root, others will. Dandelions are sturdy and good at thriving.

If our scenario for Sippewissett still ranks with the largely futuristic, we have realized a small personal attempt to protect the area in the solar retrofitting of our own house. We had several reasons for retrofitting and building a solar addition to the house where we have lived for more than ten years. The trees stand tall around it and the orchard which we planted has begun to bear. It is where our children have grown, and it is home. We had no desire to live anywhere else, but we felt a certain amount of pressure, as a result of our work at New Alchemy. Visitors to the farm would often complete a tour or a workshop on the bioshelters with a comment to the effect of, "This is wonderful. I suppose you live like this at home too?" It posed an awkward moment. We tended to shift uneasily and mutter something about not doing it quite yet. It began to be increasingly embarrassing to advocate and not to do. Our house had been built in the late sixties and as a result was poorly insulated and heated exclusively by natural gas. We asked architect Malcolm Wells to suggest a method for retrofitting the house and to design a solar addition and greenhouse. The extent of the project involved retrofitting the existing house by insulating the basement, roof, and windows in addition to building the new part. We had the one advantage of almost direct southern exposure on the front and we wanted the sun to do as much of the heating of the existing house and a greenhouse as possible. We thought that the electrical requirements for heat circulation should be minimal, equivalent to no more than two or three lightbulbs. As a result of Malcolm Wells' work and that of local solar designer Greg Wozena the solar retrofit has been completed and the workings are quite simple.

Against the north wall of the basement and ground level south-facing greenhouse are ten organ pipe, translucent, fiberglass fish tanks, which stretch eight feet in height from the lower basement floor to a person's waist in the first floor living-room. They are eighteen inches in diameter–a variation in shape on the traditional New Alchemy solar-algae ponds. They serve double duty as fish culture units and as primary heat-storage components. The tubes absorb solar energy during the day and release heat at night, warming the greenhouse and adding heat to the house. In them we grow tilapia, catfish, trout, as well as mussels and oysters in the one sea water tank. The basement is the other heat storage component which, to avoid heat leakage, was clad on the outside with four inches of styrofoam coated with a thin layer of stucco. The interior basement walls and the contents of the basement, including furniture, and a boat, firewood, tools, as well as what one writer tactfully called, "the flotsam and jetsam of a family of five," store heat from the greenhouse.

The solar heating is primarily passive, as the design called for thermopane tempered glass on the southern exposure to capture light and heat. The heat distribution in the living areas of the house is both active and passive. The stair wells and air vents in each of the rooms permit a passive upward flow of warm air. A ceiling fan in the living room circulates air warmed by the woodstove. In the summer the glass on the south side acts as an air accelerator, drawing in cooler outside air which, passing through the house, is drawn upstairs and exits through the north windows. We are sun-warmed in winter and cooler in the summer than we used to be. Internal insulated shutters on the inside windows reduce heat loss at night and shut out draughts completely.

The project is by no means over. Our future fantasies include, sometime down the line, photovoltaics on the roof and a solar hot water heater, to be installed in the upper part of the interior of the greenhouse. We already have the smug satisfaction of our own salad greens, tomatoes and herbs in January, and incongruous pleasures like seeing the sun glance off an icy mound of snow and dazzle an icicle even while it is raising the thermometer inside the greenhouse to seventy. Winter need not be one's sole reality during the months of hard, frozen ground. Earthy smells, flowers and fresh vegetables are with us all year, because the greenhouse is open to the living-room, not separate from it. A tangerine tree blossoms in early December, filling the house with the smell of citrus while we do battle with whiteflies and aphids, monitor the weather, and are doubly glad to see the sun.

We will be discussing a number of other designs which have incorporated bioregional criteria throughout the course of the book, many extant, others still ahead of their time. We include one more in this section because we feel it illustrates so clearly the precepts of the design within which we are working. It was done in collaboration with the architect David Sellers for the Lindisfarne Institute near Crestone, Colorado, but it is well suited to much of the rest of the arid, mountainous southwest.

Lindisfarne has an ideal location for a solar village. We have suggested a hamlet which, seen from a distance, will seem so much a part of the landscape as to be barely discernible. Unlike our low lying, sea-bound bioregion,

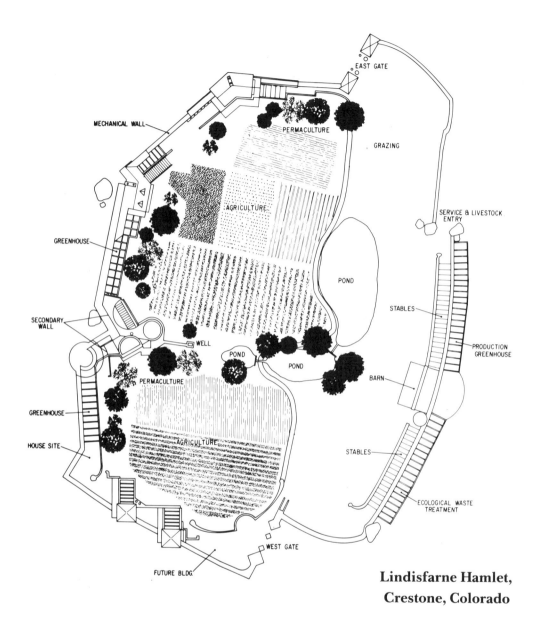

EAST GATE

MECHANICAL WALL

PERMACULTURE

GRAZING

SERVICE & LIVESTOCK
ENTRY

AGRICULTURE

GREENHOUSE

POND

STABLES

SECONDARY
WALL

WELL

PRODUCTION
GREENHOUSE

POND

POND

PERMACULTURE

BARN

GREENHOUSE

AGRICULTURE

STABLES

HOUSE SITE

ECOLOGICAL WASTE
TREATMENT

WEST GATE

FUTURE BLDG

**Lindisfarne Hamlet,
Crestone, Colorado**

in Colorado it is the power of the continental land mass and the vastness of
the sky that dominate. At the Lindisfarne site the Rockies tower over the
plain of the San Luis Valley and the ghost of the long vanished inland sea
still lingers. The sun shines almost daily, summer and winter. Behind Lindisfarne each evening, as the sun sets, it stains the snowy peaks of the Sange
de Cristo Mountains red. Water comes from the melting snow. Occasional

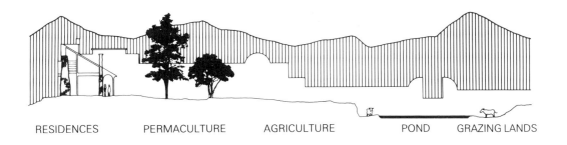

RESIDENCES   PERMACULTURE   AGRICULTURE   POND   GRAZING LANDS

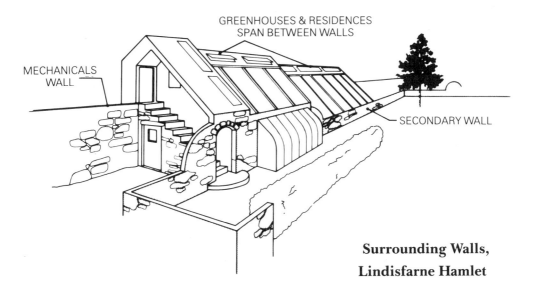

GREENHOUSES & RESIDENCES
SPAN BETWEEN WALLS

MECHANICALS
WALL

SECONDARY WALL

**Surrounding Walls,
Lindisfarne Hamlet**

streams cut through the rugged mountains creating rifts where aspen, cottonwood, and wild flowers grow—pockets that are gentler and less austere than the recurring ranges of mountains. The climate is extremely arid. It is a magnificent, but not particularly hospitable, region.

     Our plan, developed further by architect David Sellers, suggested a village grounded in the rugged terrain built to be a continuous part of that terrain. The encircling exterior walls of the hamlet are of native stone and shelter a small inner settlement comprised of clustered houses, gardens, fish ponds, livestock and fountains. The outer wall is eight to ten feet in height, insulated down the middle, and contains wiring for electricity, water and sewage piping, and fittings for the various attached buildings. It follows the ridges of the land with a rectangular shape overall, completely enclosing

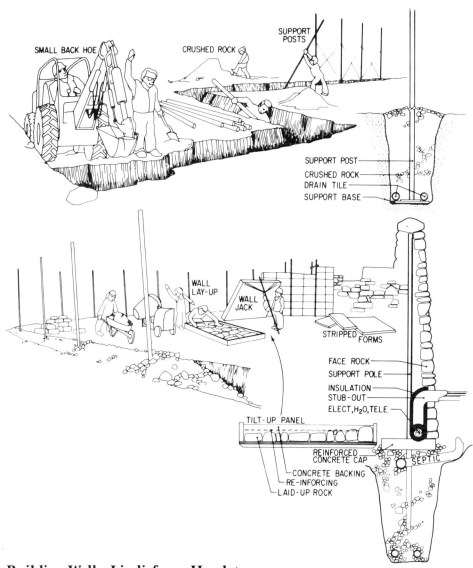

SMALL BACK HOE

CRUSHED ROCK

SUPPORT POSTS

SUPPORT POST
CRUSHED ROCK
DRAIN TILE
SUPPORT BASE

WALL LAY-UP

WALL JACK

STRIPPED FORMS

FACE ROCK
SUPPORT POLE
INSULATION
STUB-OUT
ELECT, H₂O, TELE.

TILT-UP PANEL

REINFORCED CONCRETE CAP
SEPTIC
CONCRETE BACKING
RE-INFORCING
LAID-UP ROCK

Building Walls, Lindisfarne Hamlet

the village and gardens and protecting them from cold winds and from foraging deer, cattle, and rabbits. The wall will render the hamlet almost invisible among the pinyon trees. It could be built in stages and the living components added as money and labor became available. The wall would be owned collectively by the villagers and sites for shops, houses, garages, and workplaces along the wall could be purchased. In this way the community could work its way, boot-strap style, into existence as people acquired sections of the wall.

The drawing is an early schematic of the idea. On the south face of the north wall houses are connected to each other by bioshelters. Each house has access to at least one solar greenhouse. There would be espaliered fruit trees along the house walls to the south as well as shrubs which need the extra warmth of the wall to mature at that altitude. These would create a leafy canopy to shade the houses in summer. The gardens would be sited directly in front of the bioshelter sections for maximum warnth and sunlight. Corners that receive poor light would be used for storage and equipment and all cars would be kept outside. A stream, or aqueduct, which would have a swimming pool midway, would run through the hamlet. Along the south wall on the south side a low fence barrier would retain livestock like geese, chickens, sheep, horses, and perhaps llamas. They will have their own pond fed by the stream. On the north side of the southern wall there are horse and hay barns, granaries, stalls for the sheep and llamas, and poultry yards. An orchard would be planted to give shade for the animals.

All wastes would be purified in an inexpensively constructed bioshelter of a similar design to the one suggested for Falmouth. It would be built on the outside of the enclosure that runs the full length of the south side of the south wall. Supplemental winter forage crops could be grown in a connected series of solar-algae tanks. At the opposite end of the structure commercial flower and vegetable crops would be grown for local sale. The hamlet, thus conceived, would be bioregionally sound and would provide for most of the needs of the inhabitants in terms of food, energy, shelter, and waste treatment. Like a self-contained oasis of green and warmth in that dry country, it will satisfy the needs of the inhabitants for sustenance and intimacy, yet not impinge on the silence and solemnity of the great mountains all around.

Precept Five
Projects Should Be Based on Renewable Energy Sources

We stated at the outset that it was not our intent in this book to belabor at length the ills that plague us as members of advanced technological societies. The pervasiveness of our energy dilemma, however, is impossible to overlook. Our massive dependency on non-renewable sources of energy, fossil fuels, and on nuclear energy, is one of the prime symptoms of the lack of resiliency that characterizes the developed countries at present.

Since the mid-seventies, in spite of substantial bulwarks from government subsidy and the protective vested interests of industry and the military, nuclear power has continued to fall short of the expectations of its proponents. Potentially fatally dangerous, inelegant conceptually, perenially vulnerable to sabotage, accident, and disruption, it is now obligingly proving itself far more costly than sunny forecasters first predicted for the domestication of the peaceful atom. While the fact that it is economic criterea rather than social costs that has slowed its progress to a near halt, a sign of the lingering malaise of a nearly exclusively materialistic society, it has been the efforts of environmental and humanistic groups that have forced the installation of safety requirements which have increased the overall cost of nuclear energy. Rather than go into an analysis of nuclear power, which has been expertly done in a number of books to date, we would rather confine our remarks on the subject to the observation that it is an unforgiving technology, with a timespan we cannot fully comprehend, predicated upon both human and technological infallibility. The same criticism applies equally to the development of fusion, the advocates of which are strangely evocative of those of the peaceful atom.

Before the 1973 Arab oil embargo and the subsequent entrance of OPEC onto the stage of world events, most of us accepted our consumption of fossil fuels, in the myriad forms we did so, as heedlessly and unquestioningly as we do air or water. We traveled, drove cars, powered our houses and relied on the products of the industries and agriculture than ran on fossil fuels. As students of biological principles we became aware quite early of the fallacy of depending on so narrowly based and frequently distant supplies. Our work daily brought us into contact with the workings of efficient systems and we, like many others, realized the inherent shortcomings of such a dependency relationship. In explaining our windmills and solar devices at New Alchemy prior to 1973 we would argue, often to a circle of puzzled faces, that we as an industrial culture had predicated our way of life on a contradiction–the necessity of continuing economic growth in a world that had a biological base of finite resources. To the degree that it improved our credibility OPEC did us a favor by flexing its muscles when it did in the early seventies. The culture at last understood the limits of fossil fuels during this crisis, but the cultural memory is short. People return to the earlier mind-set and use patterns without an energy crisis, forgetting that these fossil fuels are still limited and finite. Now in the eighties, it would seem redundant, but again is totally necessary to point out the inherent political and economic instability of the profligate use of fossil fuels that characterized the period from the end of World War II to 1973. We dimly understand this, as a culture, but have not changed either our domestic or foreign policy to reflect anything but a total reliance on fuel oil. Imported oil is unreliable and renders us vulnerable. Coal, although still comparatively plentiful and a potential interim possibility, raises grave environmental questions in the form of land devastation, atmospheric pollution, and acid rain. Not to be discounted are the intolerable human conditions under which the coal is mined. Natural gas, the cleanest of the fossil fuels, is not renewable and must be regarded as finite. The gradual drop in fuel consumption of fossil fuels since 1973 and the concomitant improvement in energy efficiency, although less than it might be hoped, are encouraging signs that our relationship to fossil fuels, while addictive for a period, can be cured.

As the cracks in the foundation of a society based on nuclear and/or fossil fuels become increasingly apparent, there has been a concurrent rise in interest in the possibilities for turning to renewable sources of energy to replace the behemoths we had previously tapped. It is largely a grass roots movement, moving more from the bottom up than from the top down. In their expert analysis of the energy situation in the book *Brittle Power*, Amory and Hunter Lovins list as renewable sources of energy those that harness sun, wind, water, and biomass or wastes. The transition to a society based on such renewable sources would, they point out, "eliminate the need for oil and gas wells, gathering lines, terminals, tankers, pipelines, coal trains, slurry pipelines and most bulk transmission lines."[1] They do qualify the usefulness of existing power grids for linking dispersed sources of renewable energy supplies. "Indeed," they go on, "[the power grids] eliminate the needs for the fuels themselves, oil, coal, gas, or uranium, and hence for dependence on the nations or institutions which extract, process, and sell them."[2] In addition, renewable energy is end-use oriented, and therefore ultimately more efficient. Renewable energy circumvents the clumsiness of large-scale single source strategies, which have been likened to using a bulldozer to kill a fly or a chain saw to cut butter. The particular energy need is carefully analyzed before the possibilities, which are frequently multiple, are decided upon. In many areas solar heat is best suited for space heating and has been used for this purpose with efficiency for thousands of years. There are now more than a million solar buildings in the United States. Solar hot water heaters are also growing in popularity both for domestic and industrial use. The technology for solar or photovoltaic cells for producing electricity is advancing rapidly, and their success is almost inevitable. The major problem forestalling their rapid evolution in the United States is a lack of government support—for want of which, as with the automobile industry, leadership in the field will probably go to Japan. As was the case in the development of micro-computers, the cost of solar cells is falling rapidly as both technology and materials are improved. One company is developing shingles that can be used on roofs to produce electricity. Optimistic experts in the field predict that by the late eighties new houses will be fairly common as net exporters of electricity. According to the Lovins: "The photovoltaics revolution is indeed already upon us."[3]

The fossil fuel era eclipsed, for a period, the proven effectiveness as well as the potential of windpower. Prior to the extensive rural electrification program of the thirties, many farms manufactured electricity using an on-site windmill. Interest in windmills subsequently waned to be revived only with the energy crisis in the seventies, at which time design in wind plants lagged far behind developments in other fields. Proponents and designers of windmills have moved quickly to close the gap. In the mountain passes and on the tops of California mountains "wind farms" of up to one-hundred windmills are under construction or already operational. According to *Popular Science* magazine "the wind-farm concept was kicked around through the energy crunch of the mid-seventies. Suddenly, in 1982, the idea blossomed into fact."[4] California's Pacific Gas and Electric Company buys all the power generated by these wind farms, as all utility companies are required by federal law to do. Other sites for wind farms are being evaluated in Montana, Wyoming, New Hampshire, and Hawaii. The value of such centralized power stations notwithstanding, there is, as in the case of photovoltaics, a strong case to be made for individual mills to provide on-site electricity on a household or community or farmstead level in remote areas. As with the struggle to perfect the solar cell, wind technology is improving with remarkable speed in spite of minimal government support. One member of the energy industry has pronounced wind power the closest to true commercialism of all the emerging technologies.[5]

Hydroelectricity is not a new concept, and certainly strong arguments must be made against vast constructions that change the course of great rivers and violate the integrity of entire watersheds and bioregions. Small scale hydro causes far less environmental disruption. As was the case with windmills, innumerable small dams and hydro power stations fell into disuse during the fossil fuel era. Now there is a broad revival of interest. The Lovins estimate that "Nationally over the next few decades small hydro should approach the same total capacity as existing large-scale hydroelectricity, but far more evenly distributed throughout the country."[6] Abroad, Sweden and New Zealand are actively developing small hydro. According to Lester Brown of the Worldwatch Institute in his book *Building Sustainable Societies*, in 1978 some 87,000 small-scale hydro-powered electric generators were in use in China and account for a third of the electricity produced there.[7] Other possibilities for decentralized production of electricity include solar collectors to provide heat either for direct consumption or for conversion to electricity, and heat concentration saline solar ponds which can offer heating capacity per se or produce heat convertible to electrical energy. As for energy for heating, the first and most obvious energy strategy is conservation. Once the need for energy for heating is reduced, and it can be done so drastically, like the generation of electricity, the question takes on a bioregional character. Obviously in heavily wooded areas efficient woodburners used in conjunction with a program of careful forest management and reforestation is an important and renewable strategy. In many less forested areas solar architecture combined with insulation against night heat loss has greater potential.

As for looking to renewable sources of energy to produce vehicular fuels, the most promising current solutions are gaseous hydrogen made from hydrolysis and gas produced from the processing of biomass. A note of caution is necessary here, as it is with the development of all new ideas for producing energy, even when it is not radioactive. Whereas wholesale giving over of acres of land to biomass production could encourage deforestation and soil depletion, there are at the present vast resources of waste biomass already available for conversion. In various parts of the country they include cotton gin trash, which in Texas, the Lovins estimate, is enough to run every vehicle in that state; spoiled grain, walnut shells, and rice straw in California; peach pits in Georgia; logging wastes in lumbering areas; and abroad, the straw that is now burned in the fields of France and Denmark.[8] Biomass conversion, carefully thought out, is yet another possibility for renewable energy just beginning to be seriously applied.

A satisfying account of conversion to solar is the story of Soldiers

Grove in Wisconsin. Situated on the Kickapoo River, the village grew up around a lumber mill built in 1847. As the forests on the surrounding hills were cut for timber and the land cleared for farming, the soil began to lose its ability to absorb and contain run-off, so the Kickapoo began to flood regularly after the spring rains. In some years flooding was slight, but in others it was massive and, occasionally, disastrous. After a good many unrewarding efforts to acquire government help in one form or another and repeated recurrence of flooding, community leaders, in an attempt to save the village, requested that the Federal government reallocate grant money which had previously been slated for a flood control plan. In an ingenious reversal of Mohammed and the mountain, the city leaders decided that instead of battling to keep the river away from the business district, it made more sense to move the business district and relocate it beyond the flood plain of the Kickapoo. A flood of disastrous proportions in July of 1978 finally extricated the money from the government. The people of Soldiers Grove began to take steps prerequisite to again coexisting with Kickapoo. By this time not only rising water but rising oil prices were becoming a concern to the businessmen of the village who were beginning to see, in rebuilding, a chance to restructure their downtown to reduce energy costs. A task force formed by the University of Wisconsin Extension Service recommended they incorporate maximum thermal efficiency and passive solar heating for the intended new buildings. Thus encouraged, the village hired an experienced solar architectural firm, the Hawkweed Group Limited of Osseo, Wisconsin, to design a master plan for the new downtown. The result, by the end of 1982, was the completion of the first solar heated business district in the country. According to a report by William Becker:

> Soldiers Grove shows that site-built passive solar heating systems are economical, understandable and socially/politically plausible in any new construction situation. No housing development, industrial park or other new construction should be done without an analysis of solar potential. At minimum, unless there is unavoidable physical obstruction of the sun, new buildings should be oriented properly and be designed to take advantage of direct gain. Solar should be a normal and integral part of new construction planning anywhere in the country.
>
> Because systems like those used at Soldiers Grove are practical, cost-effective and of great benefit not only to building owners and local stability, but to the nation as a whole, the burden of solar plausibility ought to be shifted. Rather than requiring local officials, solar businesses and other solar advocates to show why solar should be used, the own-

ers of planned buildings ought to be required to show why solar shouldn't be used, if they wish not to use it.

The folks at Soldiers Grove came to understand solar's benefits to the point where they passed a requirement that all new business buildings receive at least half their heating energy from the sun. Owners must seek a variance if they wish not to use solar heating. And in the one test case which has come up so far–that of the bank building–the community demonstrated that it needs to be firm about the requirement. Rather than granting the bank a variance, village officials brought in a specialist to conduct a solar-need analysis and helped the bank find an economical way to meet the requirements of the ordinance."[9]

The gradual, and as yet incomplete, shift to renewable sources of energy at New Alchemy offers an early if slightly ragged model of how such a transition can be accomplished in existing communities on a broad scale. When the Institute first moved onto the land, the only buildings were the farmhouse and the barn. The electricity for both came partly from the Pilgrim Nuclear Plant at Plymouth, Massachusetts, just north of the Cape. Parenthetically, although some of our electricity still comes from there, and that plant is one of the oldest and least safe in the country, the same combination of nuclear protest and shifting economics have brought about the cancellation of plans for a second plant–Pilgrim II. Apart from electricity, the other energy then used in the poorly insulated New Alchemy farmhouse was natural gas for the hot water heater and oil for space heating. Over the years, as ecological diversity was built into the natural systems, an analogous approach was applied to energy sources for the buildings. In all, four new solar buildings have been added, all of them bioshelters which do not require auxiliary fossil fuel heating. The Energy Education Auditorium designed by New Alchemy's Bill Smith with the help of an Apple computer was hewed out of a corner of the old dairy barn. It was constructed and insulated to make central heating unnecessary. Around the farm most of the aquaculture units have attendant windmills to pump, circulate, or aerate the water rather than using electrically-powered pumps or aeration units. A solar hot water heater has been installed on the south face of the roof of the farm house to provide domestic hot water. The house itself has been considerably insulated to reduce fuel consumption for heating. Although we have experimented with windmills and a demonstration photovoltaic unit, we are still drawing electricity from the grid and are glad it is there. New Alchemy, as the solar and wind technologies advance, one day will be energy independent. Further down the line again the Institute may contribute energy to the grid, thereby lessening the Cape's overall energy dependence on outside sources. It is a pattern bound to be repeated in a number of forms in many areas throughout the country and in many parts of the world.

The potential for renewables is not confined to replacement of what are referred to generally as conventional sources. For example, Joe Seale, Ocean Arks International's head of wind research, has invented and designed a highly efficient wind-powered ice making/refrigeration power plant for which we have been awarded fifteen patent claims. The technology grew out of a response to the very severe need for food preservation in tropical countries where spoilage is a major limiting factor in fishing and food production. Such soft technology applications of renewable energy are open to replication in innumerable ways. Many are reaching the phase of implementation. To those who still view renewable energies as inadequate to the task of nuclear and fossil fuel, it can be stated, as the Lovins phrase it, that "The methods used to forecast the path of the sun, or even next week's weather, are considerably more reliable than those which predict reactor accidents or Saudi politics."[10]

Precept Six
Design Should Be Sustainable through the Integration of Living Systems

When the late Buckminster Fuller attended the opening of the Pillow Dome, the first of New Alchemy's second generation of bioshelters in June, 1982, he inspected the building and then turned to us with a radiant and approving smile. He announced, "This is what I always wanted to see happen with my architecture–this integration with biology."

The dome is indeed integrative–beyond the combination of architecture and biology. Mr. Fuller's concept for the original geodesic domes took its inspiration from the Earth's great circle arches, from the beginning mirroring nature in the broadest sense. The integration of systems has always characterized New Alchemy's work as well. There are considerable numbers of people at New Alchemy involved in organic agriculture, in aquaculture, and in renewable energy research. Using the biological world as a model, it has been New Alchemy's intent to, whenever possible, integrate design and function. A solar pond, for example, is an aquaculture unit, a heat storage unit, and low level furnace. From the beginning we have made a conscious effort to pursue this strategy and to stretch its limits. We believe it is for this reason that an otherwise small and unorthodox institute has received the attention it has. The integrational approach is part of the fundamental policy of Ocean Arks International, also; indeed, we could not predicate any design separate from this approach.

Combining New Alchemy's ideas with those of Buckminster Fuller in the Pillow Dome represents a more complex matrix of integration than had hithertofore been achieved at the Institute. The credit for the idea belongs to J. Baldwin, then New Alchemy's soft technology expert and also the soft-technology editor of the *Whole Earth Catalog* and the *CoEvolution Quarterly*, who was a student, protegé, and devoted friend of Mr. Fuller. Responding to Mr. Baldwin's insistence that we push the bioshelter concept further and building on the information gleaned from earlier bioshelters, we were very much aware that they had been intended as prototypes. Their shortcomings had become evident. As pioneering structures they were custom designed and built, which had brought them in at high budgets even in the mid-seventies. Some of the materials had not proved completely satisfactory. The wood used in the framing was susceptible to decay at greenhouse levels of humidity and the plastic glazings were often short-lived and inadequate in terms of light tranmission. After several years of testing and observation, ideas for improvements began to surface. New Alchemy's computer technicians John Wolfe and Joe Seale framed projections on the most effective configuration and glazing possibilities for the second generation of bioshelters while the Ark staff began collaboration on improved designs for the biological components. J. Baldwin made a study of the economics and of possible production techniques for the structure. He insisted that current economics be integrated into the thinking to keep construction costs low and that it be possible that the components could be mass produced.

Drawing on his training with Bucky, he designed Bioshelter Two as a "pillow dome." The frame is a geodesic dome, thirty-one feet in diameter, made up of thin aluminum tubes which will not succumb to rot. Advances in polymer physics and materials sciences were incorporated. Computer simulations and the design team were in agreement that the glazing should be a thin membrane of Tefzel™, made up into thin, inflatable pillows. The membrane is experimental and is produced by the DuPont Company. It has a high light transmission ability and admits ultra-violet rays which increase disease resistance in plants. The inflated panels are stiff, tough, and easily manufactured. Triangular in shape, heat-sealed along the edges, they clamp onto the light frame. They are filled with argon, a heavy, inert gas, unique as an invisible insulator, which eliminates between fifteen and thirty percent of the loss of heat at night. Between the outer Tefzel ™ membranes is an inner film of very thin Tefzel ™. These pillows, triple glazed and filled with gaseous transparent insulation, represent a major breakthrough in solar design. Integrating principles of conservation with solar collection and storage, a night curtain was installed for use during the colder months to prevent reradiation of collected solar heat back to the night sky.

Inside the dome, the biological components were installed to maximize the advantages of what we had learned over the years in experimenting with semi-closed ecosystems in bioshelters. A bank of solar-algae ponds, representing the aquaculture, irrigation supply, and heat storage unit, were placed along the northern periphery of the dome, surrounding a good-sized and productive fig tree, a tenant of an earlier dome. It had been plastic wrapped to protect it during the months of construction and was the reason for resurrecting the new dome on the site of the old. It is apparently unperturbed by the arbitrary intervention in its accomodation and continues to bear prolific quantities of figs. A small central pond was installed to give visitors a chance to look at fish without peering through a layer of solar pond fiberglass and the murk of algae laden water. The central pond contains water hyacinths to demonstrate their water purifying capacity. The southern half of the dome is given over to the same kind of organic agriculture as practiced in the Ark. It has been innoculated with living soils, unlike the sterile soils of standard greenhouses. With some seasonal fluctuations it produces crops of vegetables, herbs, and flowers the year around. It is used to start seedlings in the spring and as a season extender for heat loving crops. A further integration of function is represented by the use of computers, not only in the design phase but, as was the case with both the Arks, with sensors connected to a central computer which monitor the ongoing state of the building, much as the vital functions of a patient in a hospital are monitored. The building with its semi-permeable membrane and contained living systems and integrated functions is seen, not exactly as a living entity, but as a bridge between the living and non-living worlds. It is another step in the evolution of self-reliant, solar architecture.

The mutually reinforcing advantages of integration have been widely studied at New Alchemy. One idea has spread rapidly. In 1970, while working at The Woods Hole Oceanographic Institution, we started a series of laboratory experiments in which we irrigated and fertilized lettuce with water from tanks being used to raise bullhead catfish. The lettuce fed with fish water soon completely dwarfed the control plants. Several years later at New Alchemy we repeated the experiment in field trials, using water from a densely populated tilapia fish pond. Irrigating with aquaculture water increased lettuce yields up to one-hundred and twenty percent. Now, in the 1980's, the concept of linking fish farming and agriculture is coming of age, even on large farms. In the Gila River Valley of Arizona farmers have taken to growing tilapia and catfish in their irrigation canals prior to releasing the water onto their fields of grain and alfalfa. The nitrogen rich fish wastes reduce their fertilizer needs by twenty percent, and the fish crops help pay for the expensive geothermally heated water. One Gila River Valley farmer, whose farm economics had been turned around by integrating aquaculture with agriculture, was quoted as saying, "Aquaculture is ecologically thoughtful."[1]

New Alchemy is by no means a pioneer in its use of integration in agriculture. One of the elegant applications of this principle we have seen was on some of the small farms we visited several years ago on Java in Indonesia. There was a farm, one of many that had been farmed continuously for centuries, which reflected in miniature, the major restorative processes in nature. Whereas most agriculture as a rule is short lived, lasting a few centuries at the most before the land tires and falls into disrepair, this was a farm where fertility was probably increasing each year as it had for hundreds of years–an example of a true partnership between the people and the land. All the major types of agriculture had been interwoven and balanced on one piece of land. There were trees, livestock, grains, grasses, vegetables, and fishes, but no single one of these was allowed to dominate. As significant as the disparate elements were the connecting relationships between the water/aquaculture, and land/agriculature. We in the West almost never join water and land this way, which may explain partially why our efforts in agriculture are relatively short lived. The Java farm was hilly. Although the native forest was gone, it had been replaced by a domestic forest of trees with economic and food value which protected the hillsides, farm houses and buildings, as well as the crops and fish culture below.

Water entered the farm in a relatively pure state via an aqueduct or ditch along the contour of the land. To charge it with nutrients so that it would fertilize as well as irrigate the crops, the aqueduct passed directly under the animal sheds and the household latrine. The manure enriched water was subsequently aerated by passing over a small waterfall. It then flowed between the deep channels between the crops in raised beds where it did not splash directly onto the crops, but seeped laterally into the beds.

In this way animal and human wastes were used but contamination of crops by pathogens harmful to animals or humans was minimized.

The gardens thus filtered and, to a degree, purified the water. Water neither absorbed nor lost to the garden then formed a channel where it flowed into small ponds in which fish, which require water high in nutrients, were hatched and raised. The banks of the aquaculture ponds were planted with a variety of tuberous plants. The leaves were fed to the fish and the tubers to the livestock.

The water, enriched by the fish for a second time, then flowed into rice paddies, flooding and fertilizing them. The nutrient and purification cycle was repeated. The rice filtered out the nutrients and the organic materials and the water left the paddy in a purer state. At the bottom reaches of the farm, the water entered a large communal, partially managed, pond. From time to time organic matter, including sediments, was taken from the pond and carted up to fertilize the soils on the higher reaches of the landscape.

On this Java farm integration was maximized. What was most interesting was the exacting degree to which the farmers had worked out the relationships with the patterns of balanced interdependence between the various components. Had pesticides been applied, for example, the fish, being highly sensitive to toxins, would have died and the chain of ecological relationships would have broken. The ecological integrity of the farm came from an integration of diversified, overlapping strategies mimicked from the patterns of the natural world. In more remote areas of Java, where Western ways have less impact, there is still a high level of cultural integration as well. Art and religion are as much a part of daily life as the tending of plants and animals. The sacred and the aesthetic have not been fragmented and diverted off in separate directions as is the case with much of our spiritual and artistic forms.

Precept Seven
Design Should Be Coevolutionary with the Natural World

In *The Pentagon of Power*, Louis Mumford refers to machines as defective organisms. Mumford observes: "A monotechnics based upon scientific intelligence and quantitative production, directed mainly toward economic expansion, material repletion and military superiority has taken the place of polytechnics, based primarily, as in agriculture, on the needs, aptitudes, and interests of living organisms: above all on man himself."[1] As history unfolds in the latter part of the twentieth century, one relatively untapped possibility with the greatest promise of maintaining ourselves on this planet lies in what could be seen as a partial reversal or step back from monotechnics—but with a difference. With the age of computers well upon us, the coefficient that marks our time as so different from all preceding periods is the fathomless wealth of information that is available to us. Drawing on this resource we would designate a precept that, when possible, hardware and fossil-fuel-powered machines be replaced by either information or organisms or, in a surprising number of cases, a combination of both. To do so would be one of the more sophisticated attempts at creating a working alliance between the human and natural worlds.

We have described previously a working example of this precept in discussing the bioshelter with its contained living ecosystems made up of a wide range of organisms. The building and its components are monitored on an ongoing basis so we can gather as much information as possible from this partnership of organisms and electronics. The expanded possibilities for this type of partnership are endless. Such a conjoining represents an opportunity for the extension of human intelligence and the intelligence innate to natural systems. Natural systems have a kind of thrift, a built-in recycling ability, that we have foresworn during the recent, profligate past. In a very general sense the natural world eschews waste. Unlike a number of human achievements, most natural events, even those as dramatic as volcanoes or hurricanes, are, over time, recycled or healed over. The same could not be claimed for nuclear technologies, to seize on the most obvious example of an unforgiving technology, the scars of which, if unleashed, will not heal in any time frame that has any meaning for us at all. Less long-lived but still dangerous are the thousands of chemicals, many of them toxic, with which we continue repeatedly to saturate the land, air, and water.

Nowhere is the hubris of the modern mindset and its stubbornness in failing to acknowledge a superbly worked-out, non-human resource better illustrated than in our treatment of the basic elements of soil and water. Today in the United States, as a result of mechanized contemporary agriculture, soil loss is occurring over a land greater than that encompassed by the original thirteen colonies. Globally the crisis is amplified. Deforestation and the concomitant loss of resources had already occurred on a scale that is larger than the land mass of Brazil. At the same time, in almost all urban and suburban areas, and many rural ones as well, vast amounts of organic waste materials, which easily could be converted into fertile soils, are allowed to go completely unused—an extraordinary waste of an extremely valuable resource.

At New Alchemy our soils were poor and we were interested in proving that such a process could be reversed. We undertook an experiment. The soils in the New Alchemy gardens now show the results of large-scale composting. In the early days, both the evidence of our senses and the results of soil tests proclaimed our soil to be inadequate, imbalanced, and lacking organic matter. To remedy this and build up the soil we turned to scavenging. We haunted supermarkets and school cafeterias for their spoiled food or leftovers. We scoured beaches for seaweed. We put up a sign at the town dump asking people to bring us their leaves for recycling rather than leaving them to be buried as landfill. Everything was grist to the mill of the compost pile. In the first two summers we made easily ten tons of compost. The result in three years was a healthy balanced soil that could support our organic gardens and orchards. We no longer need to make compost in such large quantity, but it is part of good gardening practice to replace what is taken from the soil in the form of the yearly harvest. Composting either in ditches between rows of raised beds and the more orthodox compost pile accomplish this—simply by providing the raw materials for the bacteria and the earthworms who are only too eager to do the job. The end product is free, and neither based on fossil fuels like boxes of chemical fertilizers nor limited to a narrow spectrum of macronutrients as are the major phosphorous, nitrogen, and potassium fertilizers. The composting process, in combining a human need and the activities of thousands of microorganisms transforms waste material that under most circumstances is regarded as having only nuisance value.

Information, particularly information in the form of microorganisms, can be used in many ways to accomplish the given ends. We had some first-hand experience with the capabilities of organisms in fields other than composting in the early days at New Alchemy as well. At that time we were looking for an inexpensive, low energy method of purifying the nitrogenous wastes in the fish tanks in order to maintain the water in a condition acceptable to the fish. Reading an aquaculture manual that had been published in the 1920's, Bill McLarney learned that it was feasible to place a calcium substrate in the ponds which would serve as a host for a strain of bacteria that feed on the nitrogenous wastes given off by the fish. Armed with this clue, we retrieved several old refrigerator liners from the Falmouth dump and aligned them along the edge of the pond. We laid a layer of clam shells—in our case on Cape Cod it was quahog shells—for the substrate, then added water hyacinths for further water purification. Using a water pumping windmill in one case and an electrical pump in another, we pumped the water from the pond through a series of purification tanks. Aeration of the water was achieved simply by placing the purification tanks above the level of the main pond and letting the water splash down several feet on its return. We found not only that the water was satisfactorily purified for the fish in this way, but the formerly toxic ammonia was tranformed by the bacteria into nitrites and nitrates which acted as a fertilizer for the algae which were the major food supply for the fish. Unexpectedly, another level of integration was achieved.

The point here is that an area an comprehensive as a landscape can in many cases be restored with a wise use of scientific information and biological tools in place of capital-intensive strategies. In ecologically reduced places, almost invariably, one finds that the soil is too porous and has lost its ability to hold moisture for long periods. A goodly portion of the planet that has been overexploited by humans and their livestock is in this state. When soils lose their spongy, water-absorbing qualities, flash floods, drought, and loss of topsoils become the norm. Lack of fresh water near the soil surface is the major stumbling block to the restorative process. A major challenge to a restorative science will be to find ways to reverse the process, to learn how porous soils can be made to hold water again.

We were confronted with just such a challenge on a coral island in the middle of the Indian Ocean. Several years ago we were asked by the Threshold Foundation, based in London, as part of their fledgling "Islands in the Sun Project", to visit, with several colleagues, an atoll in the Seychelles and to investigate the possibilities of increasing ecological diversity there. Fresh water is a major problem on coral islands. Coralline soils are notorious for the absence of surface lakes or ponds and for their inability to hold water. Rain water is normally held in household cisterns or water is pumped up from the underground lens often found under coral islands. Unfortunately, such lenses are readily depleted and invaded by salt water, which was happening on the atoll in the Seychelles.

Reflecting on the inability of the soils to hold water at the surface, we remembered a research paper we had read about a decade earlier. Several Russian scientists had wanted to find out why bogs and small ponds often formed on top of rubble-heap hills, places which under normal conditions would not hold water for long. They discovered that this was made possible through a biological process that occurs over rather long periods of time, a process they called gley formation. Gley is like a biological sealant or plastic that forms in the absence of oxygen when the carbon-to-nitrogen ratio of accumulated plant material is just right.

We were curious to find out if surface lakes could be made on coral islands through a speeded up or "quick-time" gley formation. The atoll provided a chance to experiment, to simulate a natural process specific to bogs in Russia at least. We had a large pond dug with a small backhoe and had coconuts shredded and laid six-inches deep on the pond bottom. To add nitrogen and other essential ingredients, we gathered and added another six-inch layer of shredded wild papaya, trunks, leaves, and stems—papaya being a prominent understory plant there and suited to our task. To drive out the oxygen and produce an anaerobic environment, we laid down a layer of flimsy plastic sheeting and, on top of that, six inches of coral sand. We tamped down the bottom and the sides.

The pond was given an initial infusion of well water and then we waited for the rainy season to come. The pond filled with the rains and stayed that way. It began to act as an ecological magnet for all kinds of life forms, increasing the base for diversity in the gardens and livestock. On a visit later to the island, the ornithologist Sir Peter Scott discovered that the fresh-water pond was attracting migratory birds that rarely land in the vast expanse of the Indian Ocean.

There is another side to the tale. The outstanding soil biologist, Stuart Hill of McGill University, was with us during our stay. He lamented the fact that the soils there were so alkaline that few domestic plants could be grown in them. He decided to neutralize the garden soils by adding acids from a biological source. Because compost releases organic acids just before the process is completed, he made a compost heap. When it reached the acid-releasing stage, he placed some of it, not yet ready by ordinary standards, on the gardens where they could release their acids and make the soils more suitable for growing food. It was Dr. Hill's ecological equivalent to the expensive petrochemically-based methods of balancing soils. Such experiences illustrate that there exists a vast storehouse of knowledge locked up in the insular reaches of academic and scientific institutions that

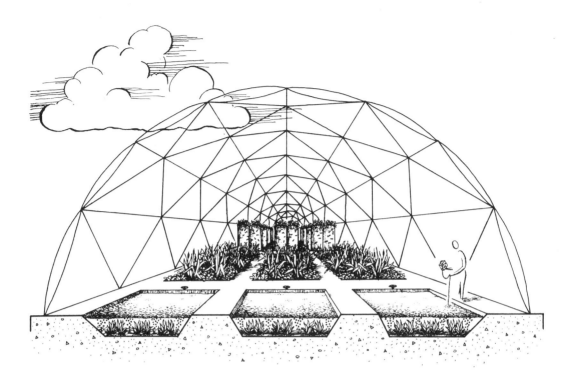

Bioshelter: Sewage Purification, Aquaculture, and Gas/Electricity Facility

can be used to form the basis of both the science and the practice of earth stewardship. The keys to the restorative process are inherent in the workings of nature.

Closer to home the concept of using natural organisms and solar energy to purify sewage is gaining acceptance. We have mentioned that a number of Cape Cod towns are interested in a new approach to sewage treatment incorporating bioshelter to enclose a warm solar-heated environment within which sewage-purifying aquatic ecosystems would be established. Such waste-treating ecosystems are being pioneered by the National Aeronautics and Space Administration (NASA) at Lucedale, Mississippi, and by Solar AquaSystems of Encinitas, California.

The Solar AquaSystems method involves preliminary treatment of the sewage and then pumping it into special ponds inside the solar greenhouse. Once inside the ponds, the sewage is aerated and bacteria, algae, microscopic animals and higher aquatic plants such as floating water hyacinths go to work removing toxins and organic matter from the water.

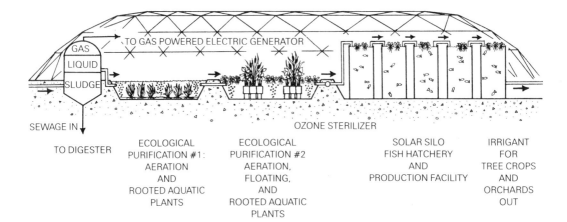

TO GAS POWERED ELECTRIC GENERATOR

GAS
LIQUID
SLUDGE

SEWAGE IN

OZONE STERILIZER

TO DIGESTER

ECOLOGICAL
PURIFICATION #1:
AERATION
AND
ROOTED AQUATIC
PLANTS

ECOLOGICAL
PURIFICATION #2
AERATION,
FLOATING,
AND
ROOTED AQUATIC
PLANTS

SOLAR SILO
FISH HATCHERY
AND
PRODUCTION FACILITY

IRRIGANT
FOR
TREE CROPS
AND
ORCHARDS
OUT

Scientists in Europe have discovered that bullrushes and sedges added to a system like this will take up harmful metals.[2]

The solar purification process we illustrate here produces by-products with commercial value. The aquatic plants can be harvested and made into high quality soil amendments or into supplemental livestock feeds. Biogas and electricity can be generated from the plants and sediments. Parts of the bioshelter can double as fish hatcheries making use of the solar heated, purified, and sterilized sewage water.

Such concepts are already proven in a number of research facilities and are being considered or tried out in a number of towns. All of these new sewage treatment plants are regarded as prototypes. However, the ecological purification elements work, and small towns, particularly those with septage wastes, would be wise to adapt bioshelter-based water purification and to propagate and market the many economically useful by-products. Treating wastes this way would no longer be a monetary drain on a community. Instead the biotechnology would create a sewage driven "farm," which would be an economic unit in its own right.

Ideas such as these are being listened to, and in some cases adopted. Their time has come. Renewable energy and the myriad skills of the organic world are gaining credibility and as energy and material costs rise will achieve the ultimate stamp of approval, economic credibility. Teamed with intelligent and sensitive monitoring and information technologies the horizon expands. Toward the end of *The Pentagon of Power* Dr. Mumford summarizes, "If we are to prevent megatechnics from further controlling and deforming every aspect of human culture, we shall be able to do so only with the aid of a radically different model derived directly, not from machines, but from living organisms and organic complexes (ecosystems). What can be known about life only through the process of living—and so is part of even the humblest of organisms—must be added to all the other aspects that can be observed, abstracted, measured."[3]

Precept Eight:
Building and Design Should Help To Heal the Planet

Like the other precepts for future design that we have enumerated so far, the one that follows is more guideline or mindset than rule. The difficulty of working with the living world and taking one's cue from the patterns discernible there is the circuitous and overlapping yet incomplete nature of what one is able to perceive of its being. Processes, structure, and functions are interwoven; everything is recycled to be born again. All is motion. All is flux. Nor is everything entirely predictable. Mutations appear. New species are born, others die and are gone forever. Gary Snyder once said that his greatest fear for the future was the irreversible diminution of the gene pool[1]—that vast resource bank of living material which gives Gaia her range of resiliency and adaptability. In attempting to formulate guidelines for thinking about the kind of design that will evolve harmoniously within the natural continuum there is a factor that has been little considered now or in the past. It now has become possible to reverse the path of ongoing destruction of the natural world which, although old or older than the dawn of agriculture, in our time has been proceeding at an unprecedented rate. We are now capable of affecting a reversal of the millennia-long tearing of the planetary fabric. We have acquired the knowledge of biology, the technology, and the potential partnership in coevolution with the organic world to begin a process of planetary healing.

By planetary healing we mean what the folksinger Pete Seeger said metaphorically when he wished for a golden thread with which he could "bind up this sorry world—with hand and heart and mind."[2] We are convinced that the equivalent of such a thread now exists in the form of some accumulated and interacting disciplines such as biology, ecology, and cybernetics, and as a result of advances in materials sciences and technology, to make large-scale restoration possible. We argue this knowing that the industrialized countries continue, as is evident in their policies and action, to behave as though they do not believe that we are all ultimately dependent on the unimpaired functioning natural world for survival. Acknowledging that, because of this, the political and sociological hurdles appear close to insurmountable, the need for such work is all the more pressing and obvious in terms of an intrinsic obligation to the natural world. It is also our duty, our obligation to provide for present and future human generations. We have elected through our work in its various forms to attempt to do so.

There are countless small precedents for such an attempt and a few major ones. The small ones are as numerous as there are well tended gardens and fields, or conscientious environmentalists and nature lovers—anyone who acts to protect and preserve any form of living species. A favorite story in ecological circles is of The Man Who Planted Trees. The man, Elzeard Bouffier, was a shepherd who lived in southern France. Starting before the first World War, when death and destruction was the norm for so much of the world, he followed his solitary path, planting trees as he went. In his wake, where hillsides had been barren, they slowly returned to forest. Springs found their way back to the surface of the ground and the trees attracted the rains again. Then people began to drift back and settle in areas they had previously had to abandon and the region again came back to life. Elzeard Bouffier managed to reverse a process so common that we tend to see it as inevitable. Instead of deforestation followed by erosion and desertification, through the efforts of one single-minded man, a range of hills in southern France was transformed from barren lands to a series of green hills that created a habitat for a rich reservoir of life forms, all contributing in the way to the overall functioning of Gaia.

Richard St. Barbe Baker was another man who planted trees. In the ninety-two years of his life (he died in 1982) he fought to save California's remaining virgin Redwoods, a campaign that was crowned with the establishment of Redwood National Park. He wrote more than thirty books, campaigned and educated tirelessly on the beauty and importance of trees, and founded a society called "Men of the Trees" to extend the work. He was convinced, as we are, that many currently deserted areas were once forested and in 1952 found evidence that much of the Sahara had once been a tropical forest. He believed that the Sahara could be reforested to support one-hundred million people. A lifelong pacifist, he believed that the world's armies should be turned to planting trees.

Wendy Campbell Purdy is a woman who plants trees. Inspired by the ideas of Richard St. Barbe Baker she began her work more than twenty years ago planting trees in Morocco. According to the publication *Manas*, four years after she had established her first planting, the trees were twelve feet high and she was able to grow wheat in the shelter which had been created by the trees due to increasing surface humidity.[3] She had a similar success, on a larger scale, in Algeria. She subsequently founded a trust called the "Tree of Life" to continue her work. Building on her already proven projects, Tree of Life plans to plant a thousand mile protective "green wall" right across Algeria in the shelter of which grains, orchards, and vegetables will be planted. Work such as this is healing the planet in every sense—an ongoing process by which the innate abilities of the human and organic worlds are inextricably interwoven and mutually enriching. Currently the life support of hundreds of millions of people is threatened by the world's expanding deserts. That this is not inevitable has been proven by the work of these dedicated tree people.

Restoration is also possible and worthwhile and rewarding on a scale far more modest than in the magnificent work of the three legends cited above. The valley below our house on Cape Cod has evolved through a number of phases. It is the inland end of what is a salt marsh half a mile closer to the sea. The area conservation officer tells us that he can remember when it was pasture for cows. When we first moved here in 1970 it was a tangle of underbrush. We installed goats for milk for our children and during the period they were growing up the valley was managed by the goats who were fairly indiscriminate and generous in their pruning methods. When we retired the goats we found that the plant range had been reduced to not much more than monocrop goldenrod. We have since installed a small vegetable, herb, and flower garden, extensively fortified against the omnipresent and omnivorous woodchuck population. Outside the garden fence we are inoculating almost at random, various meadow plants and wild flowers. If weeds like mullein or daisies appear in the garden, we transplant them. We have transplanted rose mallow from the neighboring marsh in the wetter areas. The greatest determinant of what survives is, of course, whether or not the woodchuck finds it appetizing. Even the small and almost sporadic efforts we have made so far in encouraging plant diversity, however, have been more than rewarded by the variety and number of birds that have been attracted. Far greater than in the thick of the neighboring woods, there are times, in all season, when the valley literally hums with the movement and sound of birds.

Such an experience can be replicated on a far greater scale.

Although there are few areas in the world where the primal ecological integrity has not been violated, it is our hypothesis that there is a chance that the ancient ecology lives on, but in scattered forms—in bits and pieces in various parts of the world—where it is available to be reassembled. Taking as an example the depleted shores and waters of the Mediterranean, and envisioning how magnificent they must have been before the area fell heir to its destiny as the cradle of our civilization—there are other environments around the globe analogous to that of the Mediterranean. Some of the species differ somewhat but similar life forms with comparable structural relationships exist in parts of California, Chile, Australia, Africa, and the Indian subcontinent. It might be possible that organisms gathered from such areas combined with those in the Mediterranean itself contain, in aggregate, a sufficient array of species from which to restore or recreate the ancient ecological integrity of the region. We have drawn up a pilot project to begin to tackle a project of such vast proportions, one that integrates our experience in biotechnology and the replanting ideas of Bouffier, St. Barbe Baker, and Wendy Purdy Smith.

The first step would be to create salt marshes in low-lying valleys. To do so we would install New Alchemy sail-wing windmills to pump sea water into low-lying coastal valleys. The sea water would flow by gravity back to the sea, the windmills providing a technological analogue of tidal action. The newly created salt marshes would then be planted with a variety of organisms and seeded with marine creatures collected from relic Mediterranean marshes. At this juncture ecologically-based mariculture could be undertaken to provide the restoration process with an economic base.

As the salt marsh becomes established the plan would be to plant brackish-water-tolerant plants, including the commercially important carob tree, around the edges. Many of these salt-tolerant plants would serve as an ecological beachhead for less tolerant plants on adjacent ground above. As the salt marshes start to act as catch basins for seasonal rains, this process will speed up. The marsh would begin to host a wide diversity of life forms, moving outward from the center, which in turn could trigger a more ecologically complex restoration cycle. The marsh complex would have the additional benefit of enhancing nearby marine life by acting as a nursery for many organisms that spend much of their adult lives in the sea.

We have mapped out a further restorative strategy that is more technological and would be particularly applicable to arid or impoverished areas. Bioshelters would be constructed for distilling sea water with the long range intention of nurturing young forests. The bioshelters would be approximately fifty feet in diameter and use New Alchemy's pillow dome structure. About a dozen would be pitched in a circle, like an Indian encampment. Inside the central zone of each structure would be the translucent solar tanks or solar-algae ponds to grow fish and to heat and cool buildings. During the day, relatively cool sea water would be pumped into them. The temperature differential between the water in the ponds and the air would be enough to cause the tanks to sweat fresh water down their sides onto the ground. At night the air would loose its heat to the atmosphere and the moisture-laden air within would condense on the inside of the bioshelter skin and "rain" down onto the ground inside the periphery of the building. Trees and other plants would be planted in the wet zones created by the "weeping" of the bioshelter. Once their roots were established and compost rich soils created, the protective embryo of the bioshelter could be lifted off and taken to a new site to repeat the process, leaving behind the newly liberated ecosystem. Hardy trees could be planted adjacent to this nucleus to further diversify the restoration process. Each bioshelter might be in place for two or three years before being moved to the next locale. There are many possible variations on the salt marsh and bioshelter schemes and a number of intermediate approaches. Taken together, they add up to an assembly of biotechnologies which can serve the restoration process—early catalysts in the coevolutionary process of planetary healing.

Precept Nine:
Design Should Follow a Sacred Ecology

In his essay on "Form, Substance and Difference," Gregory Bateson stated, "The individual mind is immanent but not only in the body. It is immanent also in pathways and messages outside the body; and there is a larger Mind of which the individual mind is only a sub-system. This larger mind is comparable to God and is perhaps what some people mean by God, but it is still immanent in the total interconnected social systems and planatary ecology."[1] This undifferentiated interconnectedness of the human and natural worlds in an unknowable "metapattern which connects" is what we have come to think of as sacred ecology. It is the foundation and the summation of all the preceding precepts of design. At the Solar Village Conference Keith Critchlow had reminded us: "The necessity of the sacred attitude is one of remembering–remembering the larger context of one's existence, one's duties to one's environment and to the invisible principles that regenerate life constantly." The Native American leader Philip Deere of the Muscokee tribe added: "You cannot destroy our kind without destroying nature and you cannot destroy nature without destroying the Creator." Philip Deere and his people treasure the past and present in the active manner that Keith Critchlow and the poet Annie Dillard imply. They incorporate the sacred into a living context as do many other traditional cultures whose ways have become peripheral to the dominant thrust of global events. At the Solar Village Conference the Chinese American architect Paul Sun described the world view of Feng-Shui, traces of which can still be found in modern China. Feng-Shui had its beginnings in celestial observation and profound spiritual reflection. The system translated into daily life and practice. Feng-Shui, literally, "Winds and Water," is also known by the more poetic name of Kanyu, "The canopy of heaven and the chariot of earth." It is the art of cooperating and harmonizing with nature so that a balanced life will result for the inhabitants and their descendents in a given dwelling. Feng-Shui determines where houses and other structures should be sited, as well as gardens and fields, in relation to land forms and the presence and movement of wind and water. The superstition that cloaked the ecologically sound pragmatism of Feng-Shui behind lurking threats for evil or inauspicious acts, gave it authority over uneducated farmers and villagers who might have abused the land.

Like modern solar and earth sheltered architecture and the underground designs of Malcolm Wells, Feng-Shui recommended that, when possible, a house should face a southerly direction with its back to a large hill. Like Druidic tradition, Feng-Shui specified planting, growing and harvest instructions. It would advocate, for example, that plum and date trees be planted to the south for sun, apricots to the north as they prefer cool shade, willows to the east where they create leafy shadows, and pines to the west for shade from the slanting rays of the sinking sun. Embracing the profundity of Taoist wisdom, Feng-Shui was yet totally accessible for guidance in humble matters. It was a world view that informed, directed, and gave meaning on a daily level to a stable culture.

In ancient times and from time to time since, a larger construct, one that reflected a larger reality, has held sway in the minds of cultures everywhere. The universe was the source of understanding for life and perceived as mysterious, relevant, and alive. The stars were thought to be the origin of gods. From generations over millennia who traced their courses, the sciences of astronomy, numbers, and music developed. The patterns of celestial movement were seen to be closely connected to human and universal destiny. They provided for a world view, and gave a framework in which the universe was comprehensible. Humanity had a clear place within this destiny. Almost everywhere on the planet there can still be found reminders of other ancient cultures that saw the sacred as the focus for their lives. We have been reluctant to acknowledge the superb intelligence of these ancient peoples. Their structures—Stonehenge, Glastonbury, the Pyramids, Tighuanaco, Machu Picchu, and countless other less monumental or less well-preserved remnants describe what could only have been high cultures. With all our complex structures, we have built little that will have power as compelling in as many thousands of years again. Our unquestioning acceptance of the concept of technological progress has blinded us to much of the wisdom of the past. It seems as though we have lost a subtle layer of what our Wampanoag friend called our instructions—or do not choose to remember them. Yet as our world view slowly begins to change to incorporate the new paradigm indicated by recent discoveries of science, particularly physics and biology, a continuum is becoming apparent that makes it increasingly plausible to forge links between an ecologically based cosmology and a sense of the sacred.

A concrete embodiment of this understanding is finding expression in the Chapel at Lindisfarne in Crestone, Colorado. Also called the Grail, it is a consecration of William Irwin Thompson's vision of a new harmony between nature and culture. Dr. Thompson has said, "It is not enough to raise consciousness. One must lower the spirit into the earth to embody a change in things as basic as food, shelter, and livelihood."[2] Originating in an idea of Dr. Thompson's and designed by Keith Critchlow working with Rachel Fletcher and Robert Lawlor, the Grail has neither decoration nor iconography. Its design is based on principles of sacred geometry which lie above and behind all temples from Islamic mosques, to Shinto shrines, to Gothic cathedrals. The intent, in the words of Keith Critchlow, is to bring to consciousness "the mnemonic or awakening quality that number and spatial intervals can have, to help us remember the underlying structure of the world and our relationship to it."[3] The Grail embodies a universal spirituality. It attests to and honors the highest insights of all religions. Dr. Thompson and the associates of Lindisfarne intend it as a crucible of the spirituality of the new planetary culture, in which everyday life, not technological progress, wealth or materialism, is again understood as sacred.

Paolo Soleri's Acrosanti is not surpassed by any other community in regard to a vision playing out the commitment to a sacred ecology. Based on a profoundly religious interpretation of human destiny as participatory in the spiritualization of matter, Dr. Soleri's work has been to found a city that will move humanity toward the Divine. He has written that, "The divine is practically all in the future and we, life, are responsible for its creation. The divine simulation is, or ought to be, our blueprint for creation."[4]

The city that is the manifestation of his vision is located seventy miles from Phoenix, near the town of Scottsdale, in the high desert country of Arizona. *Newsweek* magazine has called Arcosanti, "a mystical vision in concrete...probably the most important experiment undertaken in our lifetime."[5] To accomodate an intended high population density, the city is designed to be compact and three dimensional. The main structures of concrete, steel, and glass are built close to the edge of a mesa, leaving the flat land free for cultivation. The structures themselves look like sculptures, but they function like orthodox buildings. The city will eventually be self-sufficient in both energy and food. In its planning Dr. Soleri has united concepts of architecture and ecology into a concept he has termed *arcology*, which utilizes a number of physical and biological phenomena. Cars are not needed in Arcosanti, because it is a compact community. Efficient use of land will leave ninety percent of the land available for farming and wilderness. Industrial pollution is recycled and little is wasted. Acrosanti is a small controlled model, but the principles it is based on could be transplanted to more complex systems. Soleri believes, however, that the smallness of urban scale is one factor in creating livable communities.

Central to Dr. Soleri's principle of design is his theory of *the urban effect*, a socializing principal he notes exists first on a particle level. At a certain point two or more particles of physical matter begin to interact in ways other than statistical and fatal. Matter moves from behavior governed by the laws of physics into behavior which is organic or living; behavior becomes instinctive, self-conscious, mental, cultural, and spiritual. Soleri sees this urban effect as the natural enactment of human evolution, for which the intensity of urban life—as opposed to socially-isolated sparse rural settlement—is necessary. In his book *The Omega Seed*, Dr. Soleri describes the urban effect as, "the progressive interiorization, urbanization, of the mass-energy universe, initially deploying itself in space-time and eventually re-collecting itself, through the transfigurative process of evolution, into spirit."[6]

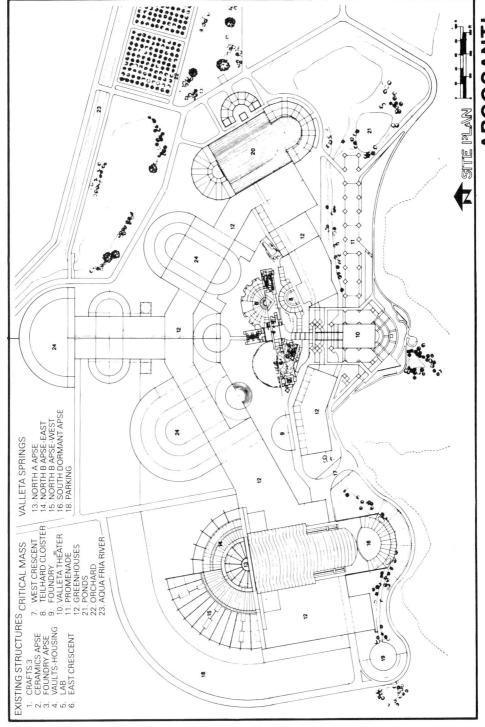

EXISTING STRUCTURES
1. CRAFTS 3
2. CERAMICS APSE
3. FOUNDRY APSE
4. VAULTS-HOUSING
5. LAB
6. EAST CRESCENT

CRITICAL MASS
7. WEST CRESCENT
8. TEILHARD CLOISTER
9. FOUNDRY
10. VALLETA THEATER
11. PROMENADE
12. GREENHOUSES
21. PONDS
22. ORCHARD
23. AQUA FRIA RIVER

VALLETA SPRINGS
13. NORTH A APSE
14. NORTH B APSE-EAST
15. NORTH B APSE-WEST
16. SOUTH DORMANT APSE
18. PARKING

SITE PLAN

ARCOSANTI

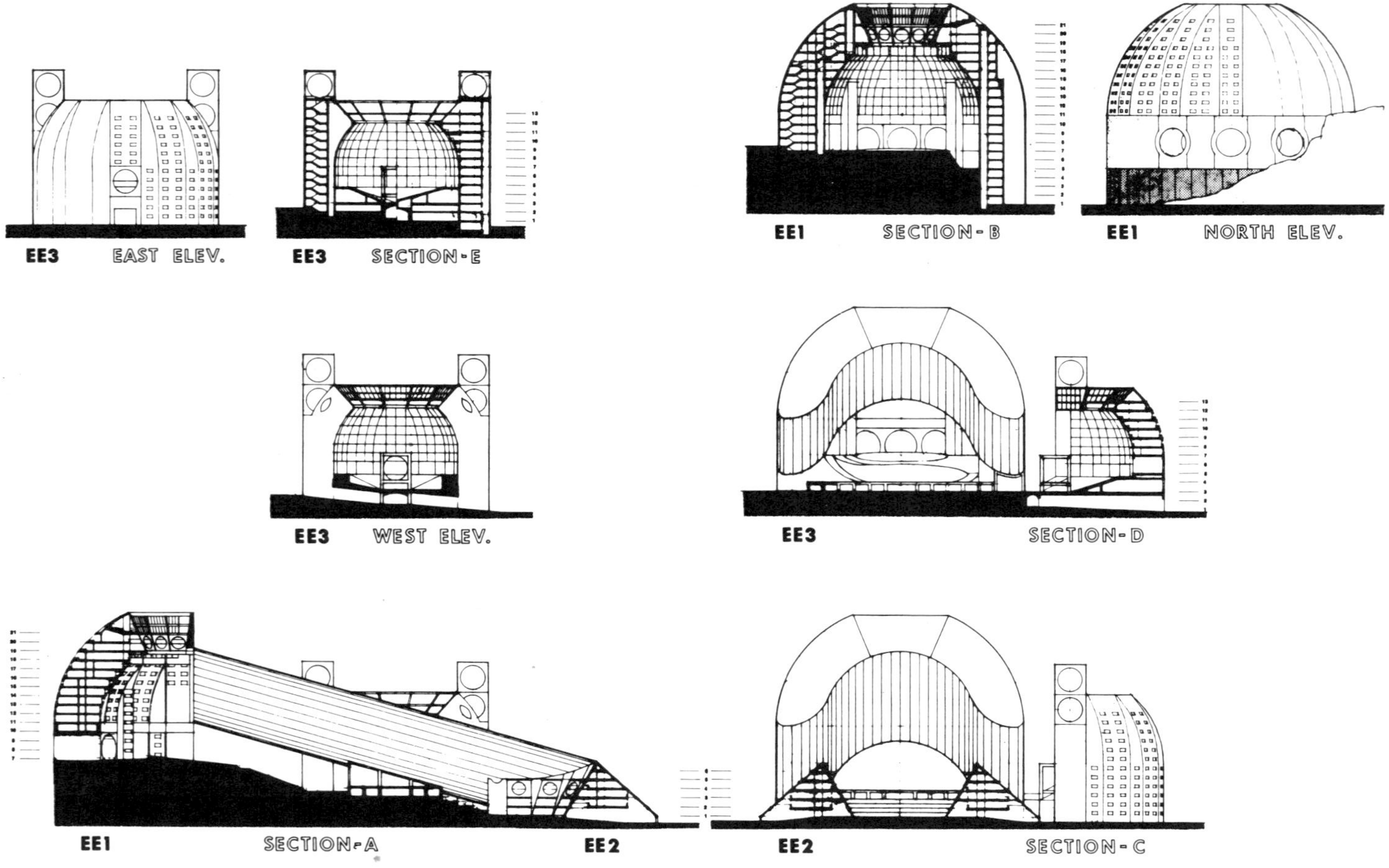

NORTH ELEV.   EE1

SECTION-B   EE1

SECTION-D   EE3

SECTION-C   EE2

SECTION-E   EE3

WEST ELEV.   EE3

EAST ELEV.   EE3

SECTION-A   EE1

EE2

EE2

EAST END, ARCOSANTI

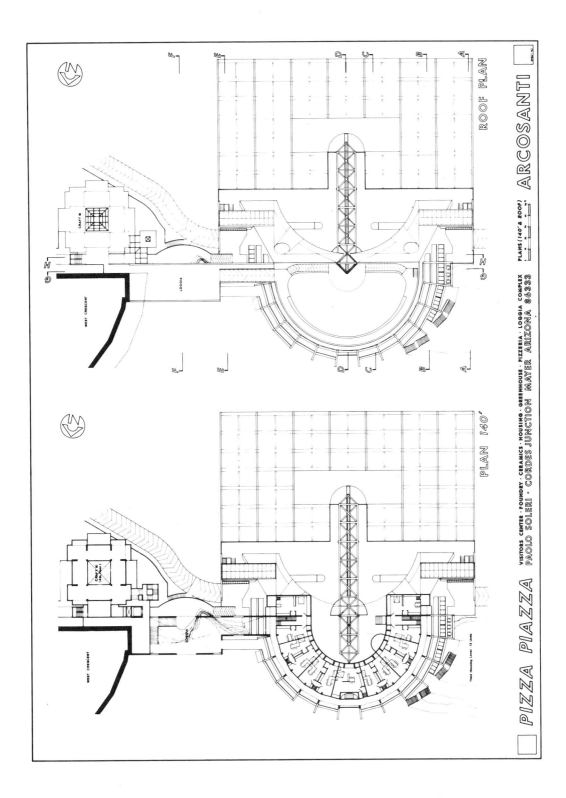

ROOF PLAN

PLAN 140'

PIZZA PIAZZA

ARCOSANTI

VISITORS CENTER · FOUNDRY · CERAMICS · HOUSING · GREENHOUSE · PIZZERIA · LOGGIA COMPLEX
PAOLO SOLERI · CORDES JUNCTION MAYER ARIZONA 86333

PLANS (140' & ROOF)

WEST CRESCENT

LOGGIA

CRAFT B

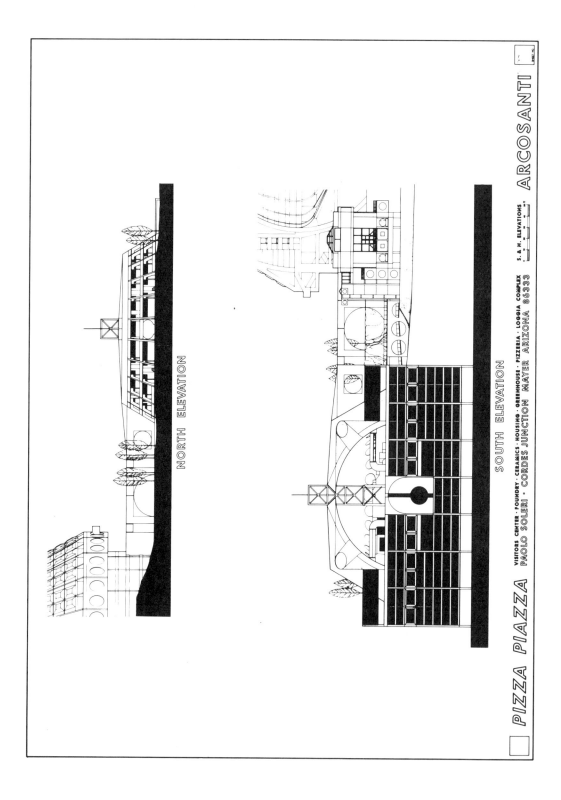

NORTH ELEVATION

SOUTH ELEVATION

PIZZA PIAZZA

VISITORS CENTER · FOUNDRY · CERAMICS · HOUSING · GREENHOUSE · PIZZERIA · LOGGIA COMPLEX
PAOLO SOLERI · CORDES JUNCTION MAYER ARIZONA 86333

S. & N. ELEVATIONS

ARCOSANTI

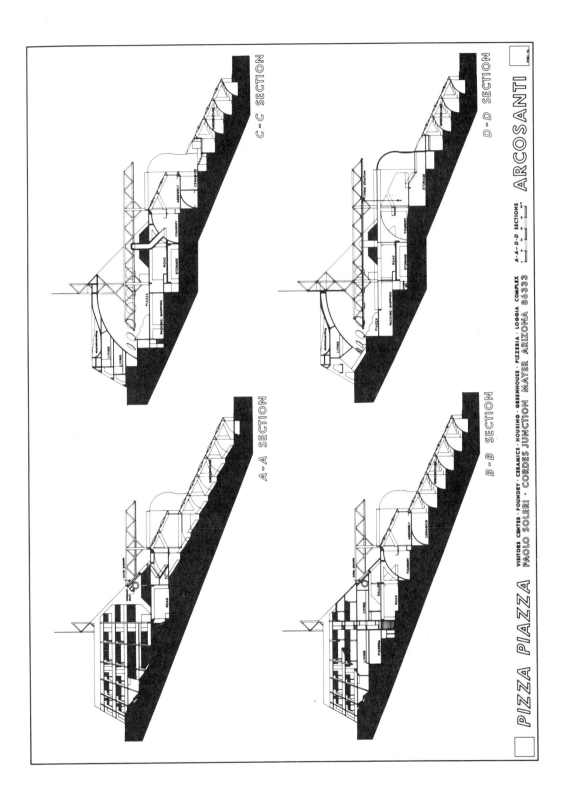

C-C SECTION

D-D SECTION

A-A SECTION

B-B SECTION

PIZZA PIAZZA   ARCOSANTI

VISITORS CENTER · FOUNDRY · CERAMICS · HOUSING · GREENHOUSE · PIZZERIA · LOGGIA COMPLEX

PAOLO SOLERI · CORDES JUNCTION MAYER ARIZONA 86333

A-A-D-D SECTIONS

LASER SPIRE

ROOF GARDEN AND HYDROPONICS

PRISMATIC ROSE WINDOW

TRIFORUM BALCONY

MUSEUM

CATHEDRAL WORKS

GIFT SHOP

BIOSHELTER

WORK

CLOISTER

THEATER

FOUNTAIN

VIEW INTO SACRISTY

DANCE & MUSIC SPACES

Cathedral Bioshelter: St. John the Divine, New York City

Almost a continent away from Arcosanti in New York City, The Cathedral Church of St. John the Divine, the largest Gothic structure in the world, has enfolded ecology into its expansive program. The Cathedral is rivaled in size and splendor only by breadth of the mission that has been forged for it by a succession of farsighted men. Under the present leadership of the Right Reverend Paul Moore, Bishop of New York, and the Very

Cathedral Bioshelter, South Transcept

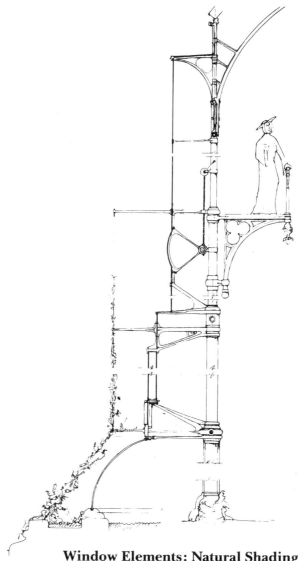

**Window Elements: Natural Shading, Glazing,
Enameled Steel and Cast Iron**

Reverend James Paul Morton, the Dean of the Cathedral, St. John the Divine is pursuing a course that honors the tradition inherent to its history and its architecture. In its time, the medieval cathedral was the center of its community, administering to all aspects of human life. Quoting Dean Morton, "Education, healing, the guilds, the arts, the market were all tied to the

Cathedral. It was the symbol of the perfection of urban life."[7] Accordingly, at St. John the Divine the arts, crafts, world peace, social justice, and ecological concerns are all part of the fabric of Cathedral life.

The Cathedral has in residence its own drama, dance, and music groups, and is, in addition, frequent host to performances by innumerable other groups, from the internationally known to the dedicated amateur. These people use the church as much more than space—they are invited to make a contribution to a community integrating the sacred and secular.

The medieval past is honored at the Cathedral, and so is the dawn of the solar age. We have proposed that the Cathedral be solar heated, as the cost of heating it is growing rapidly. Our idea is to replace the copper sheathing on the southern face of the existing six-hundred foot long roof with glass. Such a roof top greenhouse is to trap warm air which would be ducted down into the subterranean vaults of the Cathedral for later use in heating. Our plan also calls for the interior of the roof area to be used for the mass propagation of fruit, nut, and ornamental trees, which could be used by millions to help reforest New York.

The architect David Sellers grafted our ideas onto a new architectural form for the Cathedral. He has proposed that the south transept, which was never built, be redesigned as a Gothic bioshelter. It is named the René Dubos bioshelter, honoring a man who fused Christian tradition and ecological thought. David Sellers designed the southern transept with a glazed roof through which solar heat is ducted to heat the nave. Sellers' designs place solar hot water collectors on the existing south roof to heat water to be stored in a vast chamber under the crossing. In this way summer sun would be used for winter heating. The bioshelter design expresses a re-emerging relationship between Christianity and ecology. The chapel contains a garden comprised of an ecosystem specifically adapted to the Cathedral's space and climate.

As the stones are being cut for these towers, under construction again after a fifty-year lull, St. John the Divine grows daily closer to its own ideal of a Cathedral. It will be a statement in stone embracing past and future, serving the people of the Diocese of New York and of the world.

In creating sacred structures or communities which embrace ecological principles, projects like the Grail, Arcosanti, and St. John the Divine are affirming an emerging cosmology evocative of the intuitions of mystics and ancient peoples. So vast a vision may seem far removed from considerations of biological design and solar architecture, from growing food or recycling waste, but it is this vision that informs our work. Working within the framework of a sacred ecology, often at the most pragmatic levels, to implement the changes that would foster a gentler world, we may find that we are beginning, as J. B. Priestley said in *Literature and Western Man*, "To think and feel and behave as if our society were already beginning to be contained by religion—as if we were finding our way home again in the universe."[8]

Chapter Four

Redesigning Communities

1. History and Bioregion

The evolution of a given settlement or community is constrained by its location and history as well as by conflicting existing forces. There may be inherited limitations, but like a skeleton, they are the basis for structure, movement, and future action. From a biological perspective, that which has been inherited is the ground against which ecological ideas are superimposed. When most of the buildings in a neighborhood are vacant there is usually good reason for it, but the same reasons can be turned around on themselves to illuminate new opportunities. Much of the new social and economic frontier will lie in the buildings themselves and in a renewed social infrastructure surrounding their use. Function will play a large part in determining the qualities of a neighborhood. One must perceive a community in relation to its past in order to see its early riches, or lost or unexploited potential.

A good way to develop a time perspective for a community is to track down old drawings and photos and to gather as many stories as possible from long-time residents. The local historical society often holds treasure troves of information. Through the process of asembling bits and pieces of the past, a sense of the structure of a community, how it worked, and is working currently can be slowly acquired. Asking people directly what they like and dislike about a community now and in the past will help to etch the strengths and weaknesses of an area and to reveal the course for improvements. Public transportation is often a bellwether to understanding the changing economic and social relationships of an area. Another potentially fruitful exercise is to chronicle the location, type, and power output of electrical power stations over the years. Older communities have usually shifted away from local electrical generation to larger more centralized stations with greater capacity. The same pattern applies to the history of waste treatment.

Imagine for a moment a cross section of a sample coastal city, from waterfront inland to wooded shrubs. Any analysis of social patterns and their relationship to the structure of the community would require the previously mentioned historical perspective combined with contemporary aerial photographs. What would emerge is a set of relationships between residential, commercial, and manufacturing areas that directly define a given neighborhood. Historically, polluting industries and their workers were separated from middle and upper class residential areas. Commerce was usually centrally located in the center, close enough to industry to interact with it yet serving as a "barrier" shielding "better neighborhoods" from the roughness of industry and poorer areas. Industry, commerce, merchandising, and housing grew in relation to each other through a mix of practical and social influences. These earlier patterns are currently being upset in some areas as people begin to reinhabit warehouse districts, preferring to live there, and turning away from the sterility of very new neighborhoods.

A recent determinant of settlement is the interstate highway system which, since the 1950s, has been a dominant factor almost everywhere. In many instances the centers of towns have shifted to the outskirts. The interchange town and communities that have grown up are completely dependent upon the car and most probably, in the long run, on relatively cheap gas.

Few of us are used to thinking about settlement in the context of the bioregion, yet in learning to do so, we can begin to understand the ways in which places like New York, New Orleans, Kansas City, or Denver, to arbitrarily choose a few examples, are different. The respective fates of these cities were, and are, refected in the characteristics of their bioregion, just as ghost towns and abandoned villages also have something to teach us about communal failures, fragile relationships, and deep changes in fashion. Walking through a ghost town, the deserted skeleton of a social system makes it possible to trace what went wrong and to imagine what can and does work in a community.

By sheer dint of bioregion and location, New York City was fated to be one of the world's great centers. Historically, other cities like Montreal and New Orleans might have been rivals, but they have never equalled New York in influence. The reaons for this can be gleaned from a list of the city's major physical and biological attributes. The initial attributes of New York

City included: i. abundant pure water; ii. a benign climate; iii. good soil and agricultural capacity; iv. the Hudson River and direct access to the Hinterlands; v. great, hardwood forests; vi. exceptional marine resources including fisheries; vii. superb harbors free of ice year round; viii. and placement at the center oof an access hub to New England, Europe, the great grain producing mid-west via the Great Lakes, the West Indies, and Southeastern United States. The combination of these made New York City America's gateway. Coal, a key resource, was close at hand in Pennsylvania. Shipbuilding and shipping developed to solidify and enhance New York's bioregional inheritance. Although Boston had many of the same attributes, it lacked a great river connecting it to interior forests and waterways, the Great Lakes, and the continent beyond. The soils in the Boston area were thinner and the climate slightly harsher. Baltimore, similarly, lacked access to the inland and was somewhat isolated by Chesapeake Bay from central shipping routes. Montreal was a frozen port in winter, and New Orleans too distant from Europe and New England. New York, whose story by sheer dint of size has a heroic quality, has had a fate much like kings of old—it began to change as much of the original basis of wealth began to erode. Stock Markets and financial communities cannot function forever in a vacuum. The Stock Market and Wall Street are like the inner grids of computers, cut off from the materials they represent. The stock yards of Illinois, the refineries of Texas, or the steel mills of Indiana have nothing directly to do with the functioning of New York City, although they are bargained for there and support the internal business of the city. Distant sources of food, energy, and materials long ago replaced regional production. The critical economic diversity of the city and its outlying region by now have been lost, which weakens the whole—the fabric of the bioregion.

New York's exhausted biological resources include the once-productive Long Island and Jersey shore commercial fisheries eliminated by pollution, habitat destruction, and over-exploitation; the indigenous shipbuilding industry; the merchant marine which has shifted principally to foreign fleets and flags; agricultural land, as development has pushed agriculture off the best soils; and the great primeval forests. In some areas new forests ringing the population centers arc growing up again, although these are not sufficient to provide local wood.

A key element to the stabilization of New York lies in renewed caring for its basic biological resources. Although the former richness can never be regained, a plurality of activity that would increase social flexibility, activity, and dynamic quality can be created. The following activities, in combination, would help return much of the resource base to the region.

1. The modification of housing and commercial developments around New York to reduce sprawl.

2. The recycling by industry of water and airborne wastes at source to improve environmental quality so that conditions are more pleasant for people living nearby, making them less inclined to want to leave the area.

3. The true modernization of industry around themes of low energy requirements, internal waste purification, and highly adaptive and flexible working environments. If these qualities are combined with a shift towards high quality products rather than over-production, industry would be much more efficient.

4. The recycling and reusage of sewage, making useable by-products from it.

5. The recycling of water for city maintenance functions.

6. The redesign of urban landscaping based on food and ecological "islands", or small wilderness areas, to develop more liveable areas.

7. The rehabilitation of waterfronts, drawing on the example of South Sea Port, to include housing, mariculture, and floating complexes in order to bring the ocean and the river back into the life of the city.

8. The restoration of salt marshes and ocean fish inshore nurseries.

9. The development of coastal mariculture on a scale like that of Japan.

10. The reclamation of some of the original farm land.

11. A shift in the energy base to renewable energy sources, particularly solar, quickly and in some volume.

The gradual implementation of these ideas would restore to New York City some part of the biological riches of its original inheritance.

Cities like Tucson, Phoenix, and San Diego lack many of the biological resources that apply to New York. Their growth has been sustained by the heavy importation of water. As continued importation will not come cheaply, much of their future depends on re-using, and purifying, recycling water, and in creating industry and agriculture requiring minimal water use. Long-term viability for such cities lies in creating a profound dry lands mentality and culture in small villages similar to that of Paolo Soleri's thinking at Arcosanti in Arizona.

The Great Plains have their own characteristic beauty, destiny, and culture. Flying over the cities of the Plains one is immediately struck by the fact that they are located on rivers—Wichita on the Arkansas River, Topeka on the Kansas River, Kansas City on the Missouri River, Omaha on the Iowa River, and so on. These river veins or arteries are still essential to these agricultural cities, but to a lesser extent than they used to be. The history and bioregions of such cities have much in common. There are a number of qualities that make them viable and regulate their over-all size. Water is available in modest ~~by~~ but reliable quantities in spite of the trend, in recent years, to ~~mind~~ mine deep aquifers for fossil water to support agricultural production. They are usually situated amidst good to exceptional soils which are the real wealth of the region. The continuing fertility of the land is essential to the survival of such cities. The prevailing climate and soils are superbly adapted to grains and grasses, which are the basis for a viable livestock agriculture. Well developed air, truck, and rail transport provide connections with the rest of the country and a range of shipping modes. Finally, wind and sun are abundant seasonally and offer the potential for generating elecricity and hydrogen fuels rather than continuing to transport energy and fuels.

The major difference between New York and most of the cities of the Plains lies in the diversity of bioregional resources appropriate to them. Whereas New York lies at the "belly" of the western industrial world with access to a vast continent and has half a dozen biological resources, the soil is the principal basis of wealth for the Plains city. Due to modern agriculture methods, however, soil is eroding at a rate of tons per acre per year. It is vital that such erosion be stopped, or in twenty years the capability of the Great Plains to function as a breadbasket for much of the world will be gone forever. Water, equally vital to agriculture, must be protected. The mining of the aquifers should be reduced and ultimately discontinued before they are pumped dry.

This comparison of two bioregions from a biogeographic and historical perspective is intended to illustrate ways of thinking about options for the future. What have been seen as constraints, however, can still become sources for new directions. Each place has its peculiar attributes and knowing what they are and how they fit within the overall pattern of an area can provide tangible clues as to how to proceed. The Balinese, among the most sensitive and artistic of peoples, have a saying which indicates why their daily life has the precision, creativity and uplifting carefulness of an art form. It translates roughly as:

"We have no art; wc do everything as well as we can."

The ability to see an area as an entity–whether city, town, desert or mountain village–grows from learning to know a place. In a city, a good street or road map is a good place to start. In a smaller town a topographic map is an effective guide. Maps from the present and the past offer information for designing coherent living areas. Once a set of appropriate maps have been assembled, the first step is to delineate all the natural water flows, including streams and lakes, then water-related functions, like harbors, docks, and marine terminals. Using see-through tracing paper attached in layers over the map railroad courses should be charted and overlaid over the water courses, then major road traffic arteries and airports superimposed. From this material, patterns that indicate relationships will begin to emerge and give a sense of what is linked to what in a given area.

The next step is to prepare a biological map that includes the land, soil and, to some extent, the vegetation. All green zones including parks, open spaces, tree-lined avenues, wooded ravines and forests should be plotted. Such biotic arteries will probably be spotty and discontinuous. The structural patterns created previously can be superimposed on the green zone maps. It is very likely that transportation arteries cut through the topographic and natural connections. With the information thus assembled in map form it is then possible to begin to consider the elements that constitute the raw materials of design.

2. Purification and Recycling

> "It is the property of water that it constitutes the vital humor of this arid earth."[1]
>
> Leonardo da Vinci

Under every city there is a dark and hidden Venice, but we no longer celebrate our waterways out in the open. In its myriad cycles water is the source of all life, but when we, in industrial societies, harness it for our use in plumbing and sewage we keep it underground, in pipes as part of a system that is efficient for the user but displaces the problem to a distant site. Here it becomes either a source of pollution or demands costly and energy-intensive purification. Large sums of public money are spent to keep waste out of sight and out of mind.

In the recent past the clinical treatment of water has resulted in immense health benefits. Such practices as treating waste with pathogen-killing poisons like chlorine; separate waste treatment; and distant disposal have eliminated or reduced water-borne diseases in many areas. The development of solar technologies and a sophisticated understanding of the role of organisms indicate that it is time that the use of water again be im-

Fountain and Park Irrigation Scheme

proved. It can be allowed to resurface from underground pipes, be exposed to solar energy, and purified within the community. Water could become a very visible part of the fabric of architecture, settlements, and the landscape. Where we have done so in our work it is an integral and pleasing component of design. Pausing for a moment to recall da Vinci's description of water, we intend to take our cue from him and go beyond present uses of water to imagine ways, within an urban context, that we can again honor it as "the vital humor."

How greatly it would add to an urban setting if every urban block were to have a fountain. It would break the heat of a midsummer's day and cast a soothing spell on the quality of the surroundings. A fountain could be designed so that the lowest level is deep enough for children to play in and have a good splash, while upper tiers on the upwind side are relatively free of spray. There can be a central pool for purifying fountain water con-

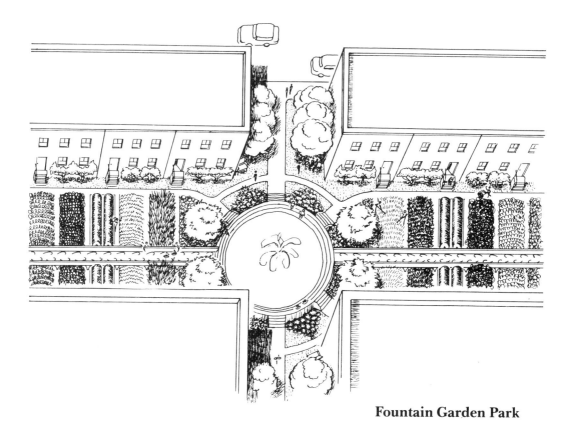

Fountain Garden Park

taining aquatic plants, snails and goldfish floating in containers. The fountain can be powered by solar cells that drive the pump which lifts the water into the air. The brighter and hotter the sun, the higher and faster the plume. It would be still at night. In cold or temperate climates the fountain water could be heated by a solar collector, mounted and covered, next to the south-facing side of the fountain which would extend the season for the fountain into fall and allow it to begin again in late spring. It would be drained in winter. The frequency with which the water would be changed would be determined by the number of people using the fountain. Used water can be used to irrigate nearby trees and grounds.

It is possible to design a new kind of sidewalk aqueduct through which flowing water, directly exposed to air and sunlight, is purified. The interior of the aqueduct, instead of being smooth with laminar flows within, is sculpted into exquisitely shaped forms and patterns which create whirlpools, vortices, and gentle upwellings. These forms are described by Theodor Schwenk in his quietly extraordinary book *Sensitive Chaos*.[2] His

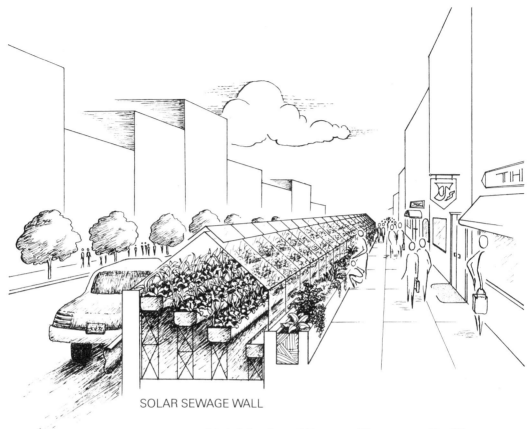

SOLAR SEWAGE WALL

Neighborhood Sewage Treatment Facility

designs have been applied as flow forms by scientists at Emerson College in England, who discovered the water purifying qualities of the forms. These forms are so beautiful that the aqueducts are functional art forms. Water that has passed through the flow forms could be used to irrigate lawns and sidewalk gardens and in this way fountain water could be transported to its end use. Such aqueducts can redirect storm and run-off water to storage lagoons, and serve other useful and non-destructive purposes. Perhaps children and adults would race model boats in the aqueduct and sidewalk regattas become the rage.

As we stated in the last chapter, new solar and biotechnologies are transforming waste treatment locally and using the products economically and ecologically in the community. We have been considering a sewage treatment idea which creates structural elements that run along streets. The Solar Sewage Wall is a long, thin, greenhouse-like structure with the

SEWAGE IN

PURE WATER OUT

← STERILIZATION

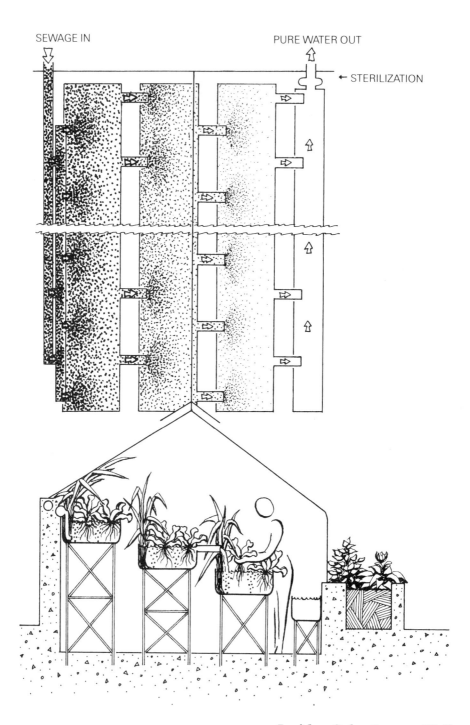

Inside a Solar Sewage Wall

south wall and ceiling translucent to sunlight. It creates a barrier that separates pedestrians from traffic. The north wall is a dark painted brick, block, or concrete mass capable of storing some of the radiant heat trapped in the structure. As currently conceived, the Solar Sewage Wall is continuous, running parallel to a sidewalk. Inside, at the intake end, is a sewage grinder and an ozone sterilization unit which kills pathogens or disease organisms. The main part of the interior is comprised of several tiers of shallow fiberglass channels or streams, sloped in steps, so that the sewage water flows back and forth and downward through the facility. The water enters at one end of the upper channel at the intake end, then splashes down to the lower channel where it flows back again to the outlet end. Finally, it is discharged, pure, into low solar water silos for subsequent reuse. Aquatic plants like water hyacinths clean up the pulverized and sterilized sewage effluent. The aquatic plants are cropped and fed to poultry or composted. By the time the water leaves the Solar Sewage Wall it is ready for any kind of reuse.

The Solar Sewage Wall would get quite hot when the sun is shining brightly, especially in summer. It is likely that at such times staff would need to adjust staggered hours, perhaps by working from 4:00 A.M. to 10:00 A.M. for part of the year. The plants and many of the microorganisms thrive on heat, however, and biological purification is optimized in water temperatures that approach 100° F. Over-heating can be prevented by venting and by training shade plants like grape vines up the outside of the south wall.

Although many European cities have turned to utilizing garbage as an energy source, burning it to generate electricity, and many Third World countries reprocess animal manures, we have been slow in North America to move in these directions. Acceptance of the necessity to recycle waste may be slow in coming. Moving waste treatment onto the block in modern solar facilities could begin to change attitudes toward waste so that sewage will be seen for the valuable renewable resource it is. The recycling process is capable of supporting a whole substructure of employment fields and diverse economic activities which contrast sharply with the expensive computer-operated chemical sewage plants remotely sited from contemporary communities.

Water can be used far more creatively than we do at present. In what we call the Venice Project, we have proposed to take one of every five city cross-blocks (or one in ten), rip up the streets, and replace them with artificial lakes or canals that would become block-long pools for swimming and bathing with shallow areas for children on one side of the swimming and bathing canal. Along another side there would be a covered channel similar

Lake-in-the-City Scheme

to the Solar Sewage Wall. Canal water will be pumped into one end of the wider central channel. As it flows toward the opposite end before re-entering the canal it will be heated and purified with the same kinds of plants and microorganisms used in the Solar Sewage Wall. Such a scheme could offer city dwellers three-season bathing and swimming. The climate of a city implementing such a scheme would be improved and both day-time overheating and night-time cooling would be lessened. In the winter, canals would be drained down to an inch or two, to make neighborhood figure skating and hockey rinks. Shops, small markets, restaurants, health food stores and sidewalk cafes could line the sun-warmed canals creating the kind of mix of commerce and recreation that Jane Jacobs advocated in *The Limits of the City* for vital and safe neighborhoods. The popularity of the fountains, pool, and skating rink at Rockefeller Center in New York and

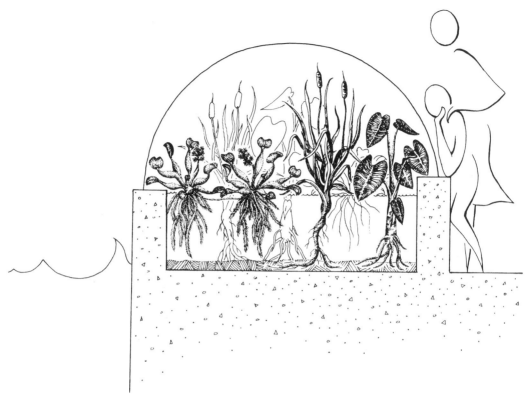

Purification and Heating for Lake-in-the-City Scheme

other municipal rinks around the country both in downtown shopping malls (Zeckendorf Plaza in Denver, Hilltop Shopping Mall in Richmond, California) indicates that like movie theaters, skating rinks are places where people congregate and should be combined with merchant development.

3. Soil

Neighborhood gardens, air purifying hedgerows, urban orchards, bioshelters, container gardening, parks, and tree nurseries, as well as the purification, recycling, and imaginative use of water are all integral to redeveloping communities. The growth of any kind of vegetation is dependent on first-class soils. A soil building program is at the core of much urban restoration. Apart from the city of Milwaukee which manufactures its sewage sludge into a sterilized fertilizer, Milorganite™, there are few other places

Drum Composter

where serious attempts are being made to utilize organic wastes. Yet in urban areas a soil building program is fundamental to urban restoration if it is to be viable. Were cities to begin a program of composting organic wastes the soil could be used for container and private gardening, neighborhood gardens, urban orchards, bioshelters, parks and tree nurseries, all of which are integral to redeveloping communities and all of which are dependent on first-class soils. Current methods of managing organic wastes on a town and city level in this country are heedlessly negligent. Mixed with other garbage, organic materials are shipped off at great expense to rot in unsightly landfills. The same materials could be separated at source and composted on a community or block or city level. All that is required are organic materials, a little moisture and some bacterial innoculants. Composting on a block scale would be ideal. It could be done in large, slowly rotating drum composters. Drum composting is hygienic and eliminates rats, mice, racoons, skunks, flies, and most garbage odors from compost in the making. The composter itself looks neat and streamlined, yet once inoculated with the right microorganisms, it is easy to operate. The drums are in sections and can be filled daily so that composting can be done in batches. The composter can be designed to absorb solar energy with a heat-absorbing-winter "blanket" and to be cooled with a reflective cover in summer.

The compost thus produced can be sieved, bagged, and marketed. Potential markets include local greening, gardening, and urban reforestation programs as well as a growing number of organic gardeners. Through processing and composting garbage, sanitation removal costs would be much reduced—especially when other recyclable dry materials, like bottles, are handled at the same source. There may come a time when compost and soil production in cities becomes an important business. Neighborhoods would find it to their advantage to establish composting, soil making, and solar sewage treatment areas and to lease these activities to neighborhood-owned corporations or private companies.

4. Working with Existing Structures

There is a science to working with existing forms and structures. It is comprised of a peculiar mixture of theory, research, and practicality—a science of "found objects." It does not attempt to build from scratch, but takes what exists and works to transform it to something useful or relevant. The French anthropologist Claude Lévi-Strauss has described it as bricolage. Practitioners are bricoleurs, which translates rather clumsily as "enlightened tinkers with what is at hand." In an age of increasing scarcity, such a person is potentially a kind of hero, someone who can see with different eyes and utilize available resources. A lack or problem is not seen only as a burden, but an opportunity. A bricoleur can see what was, is, and can be as a splendid continuum—one that must come full circle. Whereas most developers destroy before rebuilding, restorationists rebuild to recapture former glories, and designers prefer a clean slate, the bricoleur works from the assumption that the true potential of a house, a block, a whole town, or any other existing area, has scarcely been tapped. The most humble objects can be transformed. There is a human dimension to bricolage. Reflecting a responsibility to maintain continuity, the bricoleur tries to listen and to identify the voice of a place as expressed through its history and inhabitants. To transform is not to inoculate with misplaced status. When restoration leads to social displacement, it has failed. The benefits of restoration must accrue first to residents of the immediate community and subsequently to others who are attracted by the change.

Whether bricoleur or restorer, the usual way to begin renovation of an existing structure is by insulating to reduce reliance on fuel consumption for heating and cooling. To insulate is not to block out light, however. In almost any house, factory, or office, or other building, sky lighting should be first considered to create soft, beautiful, changing light. Light shafts can be elaborate or can be like simple boxes lined with reflective material like aluminum foil.

Once a structure has been tightened up and the lighting established, the next logical step is to add passive solar elements. This need not be complex. Most often it will consist of the addition of a sky light and a window greenhouse to the south side of a building. Both provide some solar gain, but the psychological effect created is important as well. The structure will become more alive, responding to clouds, rain, sun, and the changing diurnal light cycles. Solar conversion can take a number of forms. The building and site are the primary constraints on what these will be.

A purely architectural strategy for absorbing and retaining solar heat is a trombe wall built onto the exterior of the existing brick, stone, or masonry, south wall. A trombe wall is made up of a transparent wall added adjacent to the original wall which acts as a heat absorber. Sun-warmed air, heated between the walls, rises and enters the building through vents at the top as cool air is sucked into the trombe wall via lower vents. The day cycle heats the house. Whereas in winter vents are closed at night, in summer the cycle is reversed and the trombe wall vents are closed during the day when hot air is exhausted straight out the top. At night exterior and interior venting are opened to cool the building. Here at Falmouth the offices of the local paper, the *Enterprise*, receive partial heat from a trombe wall.

Another possibility for solar conversion is the addition of a passive solar greenhouse placed on the most southerly oriented side of a house or building. Generally this represents the integration of a biological element, although a solarium without plants is an option. A greenhouse addition will heat the adjacent house during the middle of the day, but cools very rapidly at night. Assuming there are plants in the greenhouse, some heat from the house will have to be bled into the greenhouse. In some instances attached greenhouses have led to greater overall fuel consumption but there are several ways to avoid that problem. We, of course, favor the solar-

STANDARD

SKYLIGHT AND WINDOW GREENHOUSES

Retrofitting with the Sun: Variations on a Theme

PASSIVE SOLAR GREENHOUSE

TROMBE SOLAR WALL

algae fish tanks developed at New Alchemy which store day-time energy for night heating. A more orthodox solution to heat loss is to install insulation or night curtains which both insulate and reflect back into the structure the long wavelengths or radiation that would normally return to the night sky. Such curtains can be mounted on tracks under the glazing with automatic or hand controls. Another possibility is a night curtain made of an inexpensive material like polyethylene suspended horizontally over the growing beds and pulled over a wire support at night. In the last century, night curtains were placed on the outside of a greenhouse and unrolled at night like a roller curtain or venetian blind. The combination of a night curtain with the fish tank heat storage has the greatest potential to conserve both money and heat. Although improving existing housing is a greater challenge than building from scratch, there are few houses that cannot be improved dramatically. Insulation, opening to the light, and adding bioshelter elements are basic ingredients.

People worry justifiably that tight buildings which have been well-insulated and have less exchange of air will feel or smell stale. Household poisons abound in many materials and cleaning products and these, in addition to cooking gases, and other smells, can make houses either dangerous or unpleasant-smelling or both. There are many ways to have both fresh air and tight houses. One obvious solution is not to use dangerous products, like some of the vinyls (new car smells of vinyl are toxic), and products like oven cleaners. Another is to install an air-to-air heat exchanger which is a simple box-like device consisting of a fan which replaces inside with outdoor air by forcing it through thin, plastic baffles. Outgoing air flows inward in thin films, giving up warmth to cool incoming air and allowing up to ninety percent of household heat to remain inside while replacing used air with fresh. A homebuilt heat exchanger is easy to make and can be done for as little as one-hundred dollars.

This engineering solution is the most common one for improving the quality of air inside houses or buildings. There are biological solutions, as well, which can purify and improve air continuously. The enjoyment to be found in the air of a forest or a meadow breeze is not illusory. Vegetation interacts rapidly and efficiently with flowing air to improve its quality and smell and these beneficial natural processes can be duplicated indoors. Placing clusters of plants near a window is an effective and simple ecological solution. The plants breathe during the day, giving off fresh oxygen and purifying the air. When there is no suitable window, a grow lamp can sup-

Household Air Purification and Temperature Modulation

port a cluster of plants. Spraying the plants with inorganic solutions to combat the pests and diseases of household plants would obviously defeat the purpose of biological purification, however.

A lidless aquarium, filled with tropical fish and floating aquatic plants which almost cover the surface of the water is a very effective air purifier. Such aquatic plants are available from tropical fish suppliers. The aquarium must be near a window or flourescent grow-lamps should be placed overhead to shine on the floating plants. During the winter lamps and window light combined will induce biological activity and effective air purification and the floating plants will add oxygen and moisture to dry, stuffy, inside air. If there is a sufficient volume of water and plants, as there is in the greenhouse in our house, the relative humidity will be increased to a point at which the temperature of a house can be kept lower without a noticeable difference. Often the inhabitants suffer fewer winter colds. One year our New Alchemy co-founder Bill McLarney took this idea a step further and moved it downstairs. He put a series of children's wading pools in his basement, suspended grow lamps from the ceiling beams, and filled the pools with fishes and plants. He called his basement McLarney's swamp. For plants there were cattails, bow rushes, arrowhead plants, and water lilies, all the emergent water plants, in fact, that he could find nearby. He then stocked his pond with tilapia from New Alchemy. These he retrieved periodically with a hook and line and ate. He maintained that watching the pools filled with exotic fish, frogs, crayfish, and minute aquatic animal life, made a northern winter, away from Costa Rica, bearable. Whenever his furnace went on, the intake air passed over the gently moving aquatic plants and the warmth of the pools filtered upstairs when the furnace was off.

Augmenting the strategies of insulation, lighting, and some form of passive solar heat gain, active solar application includes hot water and house heating. By implication, pumps, switches, and controls are involved, although, in actual fact, there is no hard and fast line between active and passive. Some passive systems use arrays of controls and some active thermosiphons are mechanically passive. Photovoltaics are active in the sense that they trigger electric currents that are subsequently channeled into useful work. Whatever the definition, active solar systems have a place in retrofitting and rebuilding. It can be impractical to think of rebuilding housing unit by unit. In neighborhoods with row or tenement housing, it makes good sense, when some general agreement can be reached, to work with

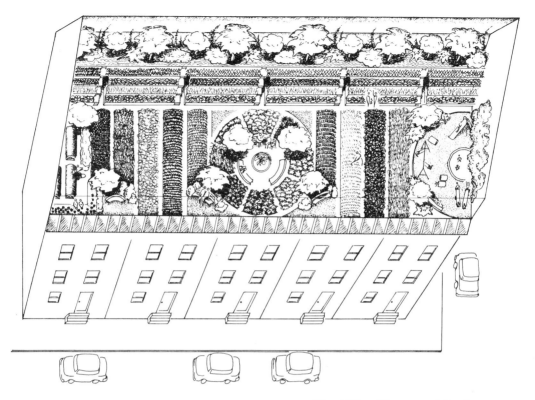

Roof Top Farm and Park

the entire block. On a block scale, available options go up and per unit costs go down. A further advantage to block scale is that there is a greater thermal and spatial mass in terms of area and heating capabilities to work with. Engineers describe this as a favorable surface-to-volume ration. It takes less energy, and fewer materials or space, to accomplish a given task.

A block of row housing can be redesigned and rebuilt to be almost exclusively solar-heated and to provide a garden environment all year. In one design the block is insulated at both ends, and on the side away from the street. The entire roof is a combination greenhouse and hot water collector. Hot water for all the houses is stored in a long, stainless steel tank that runs along the north wall for the full length of the interior of the greenhouse. It is painted black and is heated by direct solar radiation and by the collector in front of the greenhouse. The storage tank helps to heat the greenhouse and provides metered hot water to the residents. The greenhouse is connected to a common basement via ducts to prevent over-

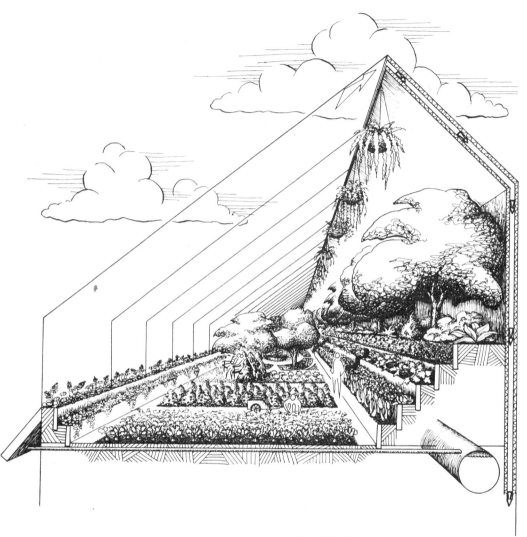

Inside Roof Top Farm and Park

heating. During the day, hot air is stored in the basement and at night it percolates upward to warm residents. The greenhouse also acts as roof insulation, retarding heat loss from the upper story ceilings. The greenhouse could serve as a commons and could be a safe and beautiful combination of roof, park, and garden. There is a strong argument for neighborhood-owned development corporations that would rebuild and operate the overall structures and support elements. In Kansas City the infrastructure for such ventures is being created through Developing Neighborhood Associations (D.N.A.s) which are self-financing organizations involved in community betterment and education. Their approach to real estate is to own houses, apartments, and commercial properties which they make available for sale or rent, thus making it possible both to rehabilitate existing facilities and to design and create new ones.

5. Biology and Architecture: the New Synthesis

In many ways architecture has failed to serve society. Since the turn of the century, lacking a broad social context, architecture has become vastly specialized—a bits and pieces business, measured in square feet. Frozen in time, much of current architecture is fad-prone, rapidly dated, and estranged from nature. In town planning functional architectural elements have become increasingly separated. People live in one kind of structure and do business in another. Commerce, marketing, trading, and government functions are all segregated, and manufacturing is located in industrial parks. The growing of food is remote from all of them. The places where most of us work and live are separated by distances that only the car can overcome.

Children are usually educated at some distance from home. Some of the saddest designs in architecture are those of schools—impersonal, often windowless, sometimes almost prison-like, buildings. The world of parents is separate from this daily fortress of the child, and the business person is separated from workers on the shop or factory floor. Few people ever come into contact with the people who grow or process their food. In a fragmented society we are all victims, intellectually and emotionally. Children do not learn to connect or to see patterns with meaning deeper than truncated parts of larger wholes. No amount of electronic information or television can alter this. In our work, we have distressingly frequently had the experience of asking children where their food comes from. After initially responding "the store!" they draw a complete blank…they cannot picture the fields, the acres, the farmers, the middlemen of agri-business. The statement that the soil is alive—made up of living matter—usually draws utter disbelief—to some people it seems like a product which can only be made rich by the addition of chemicals. For the disparate parts of society to become more reconnected, the model of nature needs to be studied. Buildings and architectural forms *can* be created in which living, manufacturing, food growing and processing, selling, banking, schooling, waste purification, energy production, religious activity, art guilds, governance and recreation are woven together on a neighborhood scale.

Restructuring the building blocks of towns and cities is now a possibility. This is partly because science and technology have reached an unprecedented juncture where centralization, specialization of function, and giantism are no longer either necessary or needed. All of the components of society, including energy, power, waste treatment, transport, and food growing can be decentralized, miniaturized, and integrated on a human scale. Such a restructuring of neighborhoods makes sound economic sense. It allows for more of the face-to-face, cashless exchanges and arrangements that make up the informal sector of economic life. It fosters the process economist Paul Hawken calls disintermediation,[3] by which he means a reduction in the number of steps and people involved in the production and distribution of goods. In many cases the most obvious place to begin this process is in the production of food. Neighborhood-grown food from greenhouses or gardens can be sold directly to the consumer by the grower. Processing, packaging, transportation, and retailing are minimized or eliminated.

In the evolving synthesis of biology and architecture a neighborhood could begin to function in a manner analogous to an organism. On the proposed block or neighborhood scale, parts become symbiotic to the whole and the basic social and physical functions work together. The workings are felt and understood by residents, who live with and operate the components.

A town or city can be looked at as made up of building blocks which are smaller communities or neighborhoods, each one similar to more remote self-organizing villages or hamlets. The communities are connected to a central zone where certain larger cultural and political events take place. The relationships are like those of the organisms to the ecosystem of which they are a part. For expression of synthesis, we look to certain earlier settlements. Rudolfsky's book *Architecture Without Architects*[4] illustrates some beautiful and profound settlement patterns. Some of the hill towns of Italy, New World pueblos, African lobi, Seripe villages on the Volta, the walled towns of the Near East, and Balinese villages all contain some of the practical and aesthetic elements and relationships which can serve us as models of functioning small communities onto which advanced solar, material, electronic and biotechnical elements could be grafted. Bioshelters could provide the connective elements. Transparent, long-lived skins on structural or tent-like tension forms, similar to the pillow dome at New Alchemy, could allow living, working, teaching, and growing to be part of an overall infrastructure. Such bioshelters could provide most of the energy needed from renewable energy sources and self heating and cooling in all seasons. They are capable of significant food production. All wastes would be within them. All of these functions reduce the need for transportation and therefore indicate smaller, simpler transportation networks.

Communities incorporating bioshelter technolgies would obviously rely on themselves for food and energy more than existing towns and cities, but they would not be islands of complete independence. Interdependencies—the exchange of fully developed foods or specialized products, crafts, or skills, still is necessary to provide spice to community life, but each neighborhood would be more complete supporting itself on the basic necessities.

Again there are helpful existing models. One of the most beautiful is the village built by the extraordinary architect Hassan Fathy in his town, New Gourna, in Egypt. The photographs in his *Architecture for the Poor*[5] are hauntingly beautiful. The drawings tell the tale of the workings of the town which, however, lacks biological elements. A walled town like Qum in Iran contains many buildings with domed roofs. One can easily imagine these contructed of transparent materials, creating a peculiarly indigenous form of bioshelter. Design for comparable communities in temperate areas would require that the whole village or neighborhood heat itself in winter yet be shaded and cooled in summer. It could have a single envelope roof, parts of which could open, or a series of roofs with a diverse mixture of glazing and vaulting to create ecosystems and provide private spaces and public places. Agriculture, the production of electricity, waste purification, education, commerce, and fabrication would become interacting elements. Most of the outer walls would be espaliered orchards and the town could be ringed by lakes and gardens.

A village or community could be created in the form of a wheel. The main spokes would be roads and the minor spokes walkway and bicycle arteries. All lead to an interior ring or hub. At the center is a lake ringed by plants and trees framed by a floral border. The village is a mixture of solid materials, transparent membranes, and mass linked to technical and biological elements. It can be thought of as a single structure with varying degrees of closure to the sky. It combines the orthodox functions of a village like housing and commerce, with a range of biological activities. The sections or wedges designated for agriculture are open to the sky in warm and hot seasons and covered by tent-like transparent envelopes during cold. Housing is compactly arranged, but each house has an individual solar courtyard that is seasonally adjustable. Waste is recycled in interconnected geodesic bioshelters. Manufacturing, maintenance, and processing are done in structures linked by solar envelope canopies, which heat the building and enclose ecosystems providing a living environment and purifying the air. Some of the agricultural zones are open for gardening by residents and there is also a section for growing aquatic foods, fishes, shellfish, floating rice, and water vegetables.

There must, of course, be the familiar structures like churches, offices, and stores, although of necessity in somewhat different form. The large aquaculture and agricultural bioshelters provide heating for all the public buildings and use the purified wastes from public buildings in nutrient cycles. In a sense the whole town with its workings and environs becomes the school. The village, thus conceived, is a truly autonomous organism, a functioning whole.

6. Growing Food in Communities

In *The Economy of Cities*, Jane Jacobs chronicled the genesis of the major agricultural innovations over the past ten-thousand years. Contrary to what is generally assumed, many of the most dramatic changes is farming including the selection of grains, the domestication of animals, mechanization of culture methods, or even recent hybrids of the Green Revolution, originated in urban centers and spread outward to the countryside. The newest agricultural development promises to be no exception. It is characterized both by an urban emphasis and ecological underpinnings. New biotechnologies, information, and biological components are being assembled into ecosystems capable of producing a diversity of foods in relatively small spaces. These can replace the fuel-powered agricultural hardware we are dependent on now and will be powered by renewable energy sources. In this way, one day, towns and cities can add farming to their repertoire of functions. Instead of farmers selling out to huge agribusiness management and coming to the cities for jobs, farmers will be able to farm in the city using new ecosystems rather than simply huge fields.

Urban agriculture will take many forms. For example, some shade trees can be replaced by urban orchards of fruit and nut trees. Sunlit walls provide convenient architectural backdrops for espaliered fruit and vine

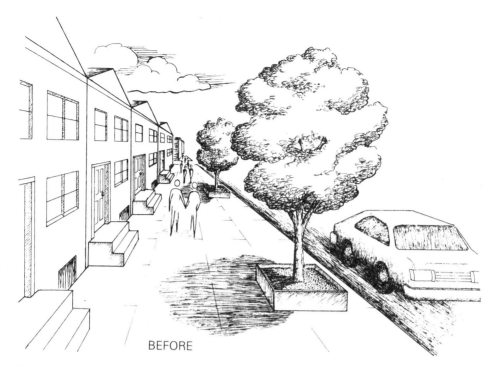

BEFORE

Sidewalk Gardening

AFTER

Vacant Lot Bioshelter Park

crops. Community gardens and urban gardening, in an experimental phase now, but growing with farmers' markets in many cities, will continue to increase. Agricultural bioshelters, which would make it possible to garden year round, could be built in vacant lots, or ringing parks. In the form of floating barges they could line harbors and sell their produce of fish, vegetables, flowers, and herbs. Old warehouses and unused factories could be converted into ecologically inspired agricultural enterprises floor by floor where fish, poultry, mushrooms, greens, vegetables, and flowers could be grown in linked, integrated cycles. Roof tops offer an unused resource for the application of bioshelter concepts or market gardens all year. Economist Paul Hawken has said that there are two fast tracks these days—energy and food. In such a context city farming has potential in the fast track. General Foods has stated that such ideas are part of the foundation for their future planning and expect to raise some foods on a large scale in the city.

It is not difficult to envision a transition from decorative plantings to street orchards and to the use of buildings as backdrops for vine and espaliered tree crop production. Both methods are well suited to many side streets. The core of a street or neighborhood can be further redesigned for an integrated mix of household food-growing and street-farming by combining raised bed vegetable and berry production, fruit and nut orchardry, aquaculture, and nurseries. The street zones could be used by the community or leased to urban farmers to grow and market crops locally. It would be up to the individual neighborhood to define and shape the forms of the gardening and farming most relevant to its indigenous ecological, economic, architectural, and cultural mix.

In heavily trafficked urban areas air is least clean the busier the street. In recent years European ecologists, particularly in Vienna,[6] have been testing species of shrubs, trees, and herbs that purify polluted air. They have found a number of plants that are effective, particularly when planted in hedgerows. In the European studies, in some cases, up to ninety percent of the lead in the air was either removed, or prevented from crossing the plant barrier. As a further bonus, street noise dropped dramatically.

A bioshelter can be introduced into an existing area quite simply. It need not be a complex structure. A three-story lean-to structure next to a building is one possibility. Such a structure will not only support agriculture, it will also provide an optimal climate for the adjacent building. Alternatively, with the addition of night curtain and fish tanks, an orthodox greenhouse can be converted into a bioshelter.

An intriguing possibility for designing a large new bioshelter is that it cover a considerable area to allow for a substantial open courtyard in the center. Possible subsystems include an urban farm around the periphery and a safe playground and a wading pool in the center. Neighborhoods sufficiently interested in the benefit to the public inherent in such a park might be willing to lease the land for the bioshelter, thereby reducing substantial land costs to the potential urban farmer.

The renaissance in urban agriculture may find its fullest flowering in the conversion of old warehouse and factories in down-at-the-heels sections of older cities and mill towns. One strategy for a badly lit multi-story warehouse or factory is to cover the roof with a full array of solar cells. Mounted at an angle to the sun, the cells would give the roof a sawtooth look. The solar cells could power indoor grow lamps, and the solar energy, transformed at about a ten percent conversion efficiency, would produce artificial sunlight inside the building. In this way the building would generate its own interior light for a variety of food growing activities. Another possibility is to replace a solid roof with a translucent one and allow natural light to permeate the whole building making six times more light available. The drawback is in heat loss to the night sky and night curtains become a necessity in winter. Should some supplemental light be required, even with the glazed roof, it could be provided by a co-generating electric windmill or by the utilities.

The interior of the converted warehouse or factory offers a wonder-

Solar/Aquatic Heating and Cooling of Office or Residential Buildings

ful skeleton for a multi-story, integrated farm that could produce ducks and chickens, fishes and shellfish, compost, mushrooms, vegetables and tropical fruits like figs, kiwi and passion fruit, and flowers. This would involve using the basement and at least three floors. The drawing of the warehouse farm company shows agriculture on four levels. The upper level, shaped something like an amphitheatre, is designed for the intensive culture of greens like lettuces and spinach. Carbon dioxide-rich air is allowed

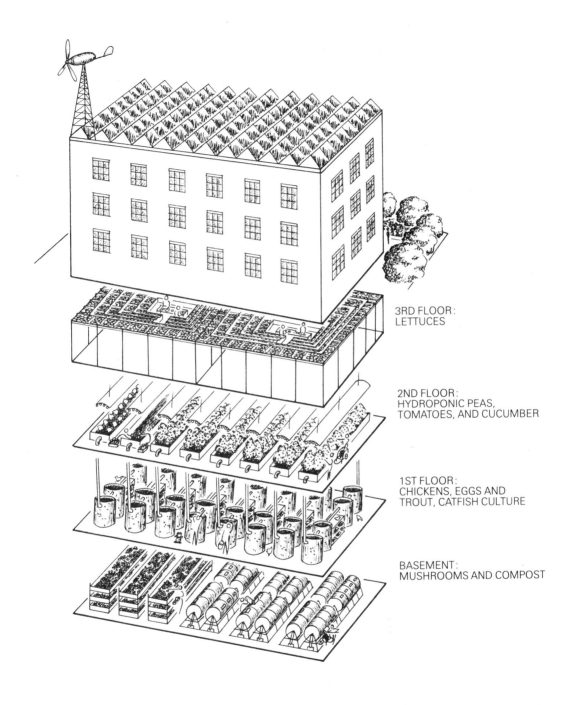

3RD FLOOR:
LETTUCES

2ND FLOOR:
HYDROPONIC PEAS,
TOMATOES, AND CUCUMBER

1ST FLOOR:
CHICKENS, EGGS AND
TROUT, CATFISH CULTURE

BASEMENT:
MUSHROOMS AND COMPOST

Warehouse Farm Company, Inner City or Suburban

to filter up through the growing beds from the lowest two levels. The next level down is for climbing crops such as peas, cucumbers, and tomatoes. They are grown in an aerated liquid solution which is pumped up from the aquaculture tanks below so that the aquaculture feeds the vine crops and in turn the root complex of the crops filters and purifies the aquaculture water in the recycling process. The aquaculture is a variation on the New Alchemy solar-algae pond theme. The fish room would double as a chicken-raising facility. Chickens would be allowed to range free on a deep litter amongst the tanks. The basement of the warehouse is used for composting of the waste from the building and for the mass culture of mushrooms. All the components in the design are integrated to mutually enhance each other. The combination of sunlight, lights, compost, and poultry produce enough heat to warm the whole complex which would require no additional heating. All wastes would be treated and recycled internally and any excess could be used or packaged as soil amendments.

The economics for converting a factory to a solar food barn are as yet conjectural. It is estimated that a one-hundred-thousand gallon aquaculture facility would be capable of producing close to a quarter of a million dollars of live retailed trout and catfish a year. The other products taken together might do as well. The limits are as yet uncertain but such a local enterprise, drawing on the community for full- and part-time staff, has the potential to reverse the present agricultural equation. The plans for such a facility have been described as reminiscent of a medieval village. The solar aspect would add the feeling of great nineteenth century English conservatories combined perhaps with an eerie, self-contained, possibly beautiful, quality of a colony in space.

In most urban areas there are flat roof tops which offer excellent opportunities for gardening and for "wild" ecological islands in the city. Roof tops can support orchardry, market gardens, greenhouses, and poultry barns, although, in this last case, all wastes must be composted immediately. An occasional building could support fish farming on the roof, but as each gallon of water weights almost ten pounds, a ten-thousand gallon facility would weigh close to fifty tons. It is essential to have a structural engineer to evaluate the strength of a roof before considering aquaculture.

In its adaptation to existing architectural forms the evolving urban agriculture will be topographical as well as ecological, as practitioners find and convert unused spaces in towns or cities. However bizarre the space, some form of agriculture can be designed to fit and operate within its con-

Bus Stop Aquaculture

straints. In integrating many different kinds of aquatic, light, soil, compost, nutrient, wind and gas cycles that are constantly changing and always interacting, any space has elements of a world in miniature.

Some forms of aquaculture or fish farming make particular sense in towns and cities. Fish are most palatable when fresh. The New Alchemy model of solar aquaculture has proved that it is both possible and cost effective to grow fish and shellfish in small spaces, using relatively little water. The translucent, cylindrical solar-algae pond provides a superb habitat for algae-based ecosystems containing cultured fish. Such tanks can be placed in almost any spot that receives direct sun, not the least of which is inside a

Sidewalk Solar Aquaculture

bioshelter. They are also well adapted to aquaculture situated at a market site making it possible to sell fresh, even live fish, on a street corner.

The New Alchemy solar tanks are among the most productive standing water aquaculture systems yet devised. Because they are solar driven, they use only small amounts of supplemental energy and therefore houses, old buildings, corner lots, parks, virtually any place lit by the sun for a good portion of the day, lends itself to fish farming. At New Alchemy in our translucent vertical tubes we have raised tilapia, trout, catfish, white amur from China, pacu from the Amazon, and mirror carp from Israel. Tilapia and catfish are the easiest to raise and most tasty. When John Hess was food editor of *The New York Times* he visited us, tried the fish and subsequently described our tilapia in *The Times* as the finest tasting of farm-raised fish he had sampled. His headline read: "Farm-Raised Fish: A Triumph for the Sensualist and the Ecologist."[7]

Any additional energy required to raise fish can be obtained from a solar panel that charges a small battery which, in turn, powers a seventy-five watt air pump, when oxygen in the pond drops to less than optimal level at night. In an urban environment, organic wastes including garbage can be pressure cooked, dried, and with additives made into pellets for fish feeds. With a little experimentation, the right food can be concocted for most fish. Generally the same feed can be consumed by poultry as well. The diet of both can be supplemented with earthworms from compost.

Perhaps the greatest advantage of this type of solar aquaculture is that it is cheap and easy to get started. A few hundred dollars in capital outlay for tanks, equipment, and tilapia or other fish, will establish a single operating pond. It is quite inexpensive to set up and experiment. When solar fish tanks are installed in a house, apartment, or office, they should be placed in a south facing room with a good sized window or skylight. The tanks absorb radiant energy and heat up during the day. A fan blows the warmth into lower rooms. Heat rises through the building and returns, cooled, to the south room, with enough time lag so that day-time heating continues through the night. During summer the cycle is a cooling one. The sun shining on the glazing creates a temperature differential between the outside and the warmed interior of the south room. As a consequence, air starts to climb and flows as an interior "breeze" out through a vent or chimney on the north side. Cooler outside air is then sucked in and moves quickly through the building. In summer the fish tanks play a beneficial role as thermal buffers, absorbing sunlight and preventing overheating of the air in the growing area which further benefits the plants. A contained ecosystem designed in this way will replace the heating and cooling functions of a furnace and air conditioner. This contained ecosystem uses no energy and has the added benefit of the fresh vegetables and fish produced—an excellent working application of trusting the partnership of organisms and information. Such an ecosystem is the structural grid for year-around urban agriculture.

7. Transportation, Power and New Shapes of Employment

The redesign of neighborhoods must include transportation, linking it to other changes. Rising transportation costs for both vehicle operation and road maintenance will be a major catalyst in altering the form of neighborhoods just as the widespread use of the car affected, over the last century, existing settlement patterns. In the present social context, transportation is a primary product, crucial to manufacturing. Automobile businesses employ one in five Americans and consume nearly forty percent of the six point seven billion barrels of oil used in the United States every year. For many of us, ownership of a car spells freedom and mobility and has come to seem a birthright. Cars will continue to be made, but our cities can't sustain them. We will use fewer as other options open, and be glad to give up the problems they cause. We can look ahead to new kinds of taxis, commuter buses and elegant, light rail, inter-city trains. We will begin to see as attractive commuter bicycles, motorcycles with climate and body protection, and small airships—extraordinary light airplanes as maneuverable in their medium as cars in theirs. Sailing ships will appear and give us alternatives to rail and plane. These may include auxiliary bioshelters on board and be self-sufficient village farms in themselves—a far cry from our luxury entertainment liners, but providing safe, interesting travel experiences. Transportation will conserve energy while increasing the quality and diversity of travel. The emphasis will be less on speed than on genuine mobility. Possibilities for travel could range from health promoting exercise to convivial transportation that offers watching scenery, dining, and studying or taking courses, like languages, electronically en route.

Changes in transportation and changes in the structure of cities are co-evolutionary. We are going to focus, as a culture, on integrating different transportation systems to create greater overall travel efficiency. In this way neighborhoods could truely come into their own, becoming neither by-passed or disconnected. A variety of new linkages and options for internal neighborhood development should begin to emerge in the next few years. Bicycles, small scooters, and walking will become increasingly popular within the neighborhood. The most desirable neighborhoods will come to be those with diversified merchants—not huge shopping malls, but merchants situated within walking distance to housing.

One of the best thinkers and designers of ecologically sound transport is Christopher Swan. His drawings indicate how transportation can evolve, from tiny electric city cars, to highly efficient taxis, to van commuter buses that provide relaxed commuting for six to nine people. In commuter buses commuters could read or watch television and have tea or coffee en route to work. Automobile development will increase in efficiency, but em-

The Sun Train and Sun Powered Roundhouse, Designed by Christopher Swan

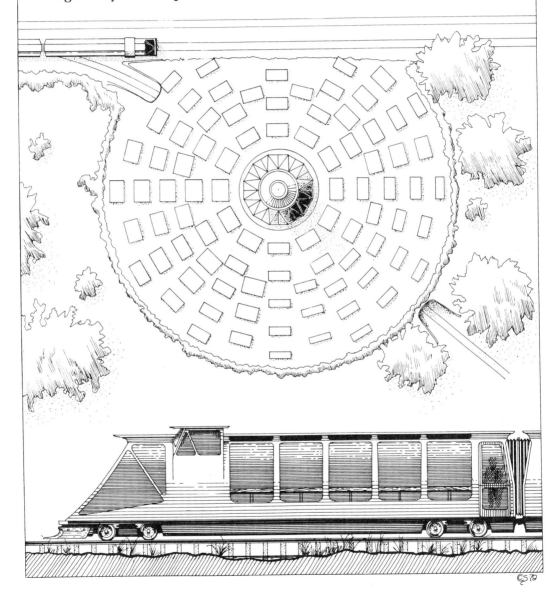

phasize preciseness of fit for each mode of transportation. Opportunities will expand for private neighborhood transportation companies, employing taxis and vans which, depending on the time of day, could shuttle four to thirty commuters as well as a cargo of supplies and materials. Other commuter transport will provide auto and bus transportation to major transportation links like subways and light rail trains.

Christopher Swan's light rail trains are the inheritors of electric trolley technologies rather than traditional heavy trains. They are designed to be powered entirely by the sun. His company, Suntrain Inc., is planning the Northwestern Pacific Railroad for travel between San Francisco and Eureka in northern California. Trains such as his represent a new and flexible dimension to mass transportation. The light rail cars can carry commuter bicycles. They could offer many kinds of cooking, solarium cars, curtains to create small compartment-like spaces, bookstore and newsstand cars, sauna and shower cars, sleepers, even movie cars. These trains are not reluctant Amtrak technology, but utilize rather the same kind of imaginative thinking and high technology that goes into Hobie Cats, Ocean Pickups, wind surfers, hang gliders, and other high performance off-shore boats and aircraft. What makes Christopher Swan's designs and ideas unique is that they are framed within concepts of ecological development for the regions through which they pass.[8]

In its early years, in the late 1800s, the bicycle was seen as a serious transportation device. For many years it came to be classified as a means of leisure, exercise, or sports equipment until the 1970s when it again began to be used as a serious transportation machine. Road capacity is designed exclusively for cars. People would use bikes more if safe pathways to ride them were designed wherever roads are built. A street network for bicycles—some streets becoming bicycle streets throughout a city—would be optimal. As for walking from place to place, developers of shopping malls, like Jim Rouse who built Harborfront in Baltimore, Faneuil Market in Boston, and South Street Seaport in New York City, proved that people like to walk when it is safe to do so, and made interesting and attractive with a mix of shops, sidewalk vendors, and restaurants. Then they prefer to park their cars and cover the ground on foot, to better enjoy the activity and the scenery. The pursuit of architectural, energy, and agricultural solutions to many urban problems simultaneously create uplifting pedestrian environments. A child, an infirm or old person, as well as a briskly walking adult should have a variety of avenues for walking. In Christopher Swan's plans for what the urban matrix could become, he has portrayed possibilities in architecture, conservation, renewable energy, bioshelters, street farming, waste recycling, parks and fountains—a combination of strategies which together contain the possibilities for the rebirth of communities. Although it is unrealistic not to be apprehensive, the future need not be frightening. Emerging from the industrial era into a post-industrial culture will include a rediscovery of what is meant when we say we truly *live* in a given place. A renewed sense of place and responsibility may emerge, echoing the lines of the poet/farmer Wendall Berry when he said:

> All my dawns cross the horizon
> and rise from underfoot.
> What I stand for
> is what I stand on.[9]

What kind of work and economic activity could grow out of the restoration of communities? As neighborhoods, towns, and cities become centers of vigorous economic life, new work will also be created. The production of compost for soil, the growing, recycling, harvesting and manufacturing of food and materials in this new world will create industries of necessary products. The jobs will be local to the community, and connect in a visible web of alliances and opportunity which used to exist in coherent communities in America but which has been lost for so many of our youngsters. Too many of us are strangers where we live, needing to invent and bring into being a "networking" that should partly be happening naturally in the communities we live in. More businesses would be at home or close to home so children would see what it is we do during the day, and in many cases be a part of it, an education in itself. Sterile streets and city blocks, normally deserted during the day while the adults are at work and the mothers are at home tending the children—or the mothers are at work and the children are together in daycare—these streets should be active with a life of shared community as the city surrounds and gives everyone a role. The community itself, made carefully by its members, would be endowed with a sense of place and beauty and its own quiet, but important identity and destiny. Instead of feeling estranged and lonely in a city which is too big and too impersonal, every living space would be part of a smaller neighborhood that would be knit into one whole by the necessity of running operations together. Bits and pieces, like The Developing Neighborhood Associations of

Kansas City are already in place working all across the country. Productive, meaningful work in the sense of the dignity of holding a job as well as the necessity of making an income is the basis of a sustainable future anywhere.

Ecological design generates new complexes and systems which make an impact on manufacturing. It has become possible to dovetail new manufacturing into communities and to integrate production, education, food production, waste treatment, housing, and the environment into an ecological whole. Almost all of the assembly and much of fabrication that today is done in large plants and factories could be decentralized and transferred to small shops. In recent years sophisticated micro-electronics and machining equipment have transformed manufacturing. Increasingly, many sophisticated machines, computers, airplanes, hang gliders, boats, and even automobiles are micro-manufactured. Christopher Swan wants to have his railroad cars manufactured in buildings which are part bioshelters along the right of way. He plans to plant forests of fast-growing trees along the route which would be cropped and converted in small factories to wood/epoxy composite materials to be used in building the railroad cars. Already aircraft, giant windmill blades, and high performance boats such as the Ocean Pickup are being built with wood/epoxy composite materials.

Micro-manufacturing can mean cottage industries which would strengthen local economies but do not break up neighborhoods. In the city of Bandung in Indonesia there is a neighborhood where many of the residents are employed by the Phillips Company, the Dutch electronics giant. In an interesting mixture of old and new industrial practices, Phillips has fostered a cottage industry in which people work at home in tiny rooms amidst their families, building parts for subsequent assembly into the various sophisticated Phillips products. It may come to be as pertinent to ask what manufacturing cannot be miniaturized and decentralized as what can be. Perhaps one day a steel foundry could be prebuilt to fit into a neighborhood.

Linked to the challenge of restructuring manufacturing into a larger social and ecological context is the need to rethink the ways in which energy is produced and used. The production of energy should be intrinsic to process—meaning that each unit, in as much as is possible, produces its own power and is designed to function around the natural oscillations of sunlight, wind, and gas production. The first step, one that is already beginning, is a heavy emphasis on reducing power needs through conservation and integrated design.

Communities have considerable potential as co-generators of power.

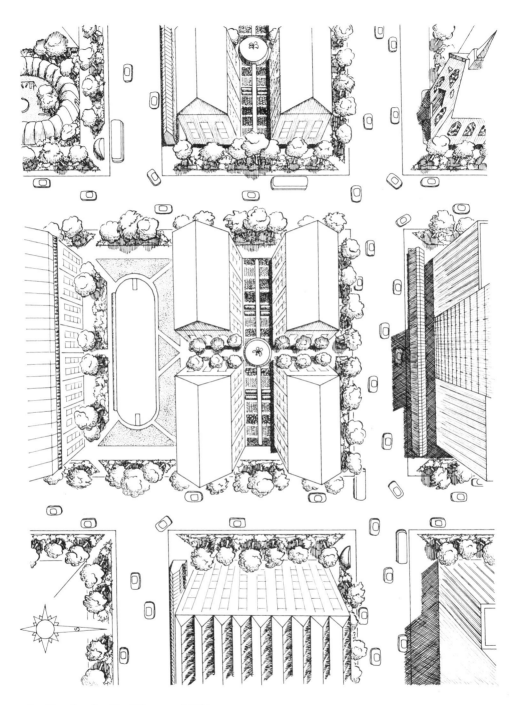

An Ecologically Planned City

In partnership with other co-generating communities, they can be linked to each other by existing power grids. As the application of renewable energy matures, the miniaturization and spreading out of production becomes increasingly possible. Initially, the utilities will need to provide back up for the whole system, but will not need to produce at peak load capacity because communities will be designed to continue to function when the grid fails. Sophisticated computer programming could provide both energy producers and users with a range of options matched to the source and amount of power being produced. Such internal self regulation will not be unpredictable and erratic. At the user end, whether house, factory, or transportation device, adaptive techniques would have to be built-in to deal with variable inputs.

As a neighborhood becomes revitalized, every activity will yield varied kinds of employment. One of the major areas will be the growing, processing, and distribution of food. Remodelling, construction, and rebuilding buildings and streets will also offer openings for many skills. Energy production and waste recycling component units will need staff. Communities could consider incorporating specialized and unique, either mass-produced or hand-crafted, components in their rebuilding, making quality, longevity, and durability hallmarks of their work. Eventually it may pay not only in personal but in economic terms to build for one's grandchildren again. Hopefully they will stand near us as we work.

As a neighborhood becomes a center of integrated activity, there will be virtually unprecedented opportunities for young people. Adults as well will have an increasingly balanced talk-do ratio while learning new jobs, skills, and sensibilities. Educators might have to rethink what is meant by learning. In a community functioning in ecological balance and with its parts exposed to full view, the patterns and cycles, natural and human, are balanced and interconnected. The community itself is the school; because it is designed after the larger workings of nature using biological precepts of design, it is like a world in miniature. Living in such a place young people may take an integral part in all that is going on and want to participate in the peaceful transition of the planet from one based on the production of goods, to one dedicated to a fulfilling life-base for all its living creatures.

Chapter Five

The Surrounding Landscape

1. Agriculture in the Past

The arguments in favor of the decentralizing and urbanising of agriculture which we offered in the last chapter are not intended to negate or belittle the importance of a revitalized and restructured rural agriculture. Rural agriculture is the basis of our planet's ability to feed large populations. The development of agriculture evolved with and predetermined the earliest settlements. Its practice reaches back into prehistory and has left an indelible imprint on the surface of the Earth. The use of fire, the domestication of animals, and the selection and growing of plants predate recorded knowledge. A human bond with the land and the daily and seasonal rounds of plants and animals span the centuries and provide an unbroken link between the earliest humans and farmers today.

Three-hundred and fifty thousand years ago Peking Man (*Homo erectus*) discovered the use of fire which was applied, carefully and artfully, to tame the landscape. In spring and late fall our ancestors lit small fires to control ecological succession, keeping meadows open and ensuring that grasses would be young and tender. The purpose of burning was to stop or slow ecological succession, thereby preventing meadows from becoming thickets and then dense woods. The big game animals preferred open areas. Low temperature fires maintained burned-over meadows which acted as magnets for the herds. The animals remained wild, but the landscape was controlled by fire.

Burning was both art and science. The skills have been passed down from generation to generation into the twentieth century. In northern Alberta, in Canada, a few of the oldest Native Americans of the Slavey, Beaver, Cree, Chipewyan tribes, and some of the Métis remember how to farm with fire.[1] In their younger years, season, weather, and frequency of burning were still understood as was the need to maintain the delicate balance between natural creation and environmental destruction. They knew that hot summer fires or too frequent burnings over-exposed the land, causing it to lose fertility, whereas burning with the melting snows or at the beginning of the rainy season contained the fires and protected the soil so that pastures remained lush. In terms of planetary evolution grasses were the most recent plants to appear. Many of them co-evolved with fire-using early peoples. Grasses, which include the plants which subsequently came to be utilized as edible grains, are highly adapted to fire. Their extensive subterranean root systems and ability to grow up quickly after cutting or burning permit them to be constantly regenerating, often at the expense of other plants.

The domestication of animals followed fire as the next major innovation in agriculture's long history. Great herds of game wandered over vast ranges and it was not always possible to maintain the hunt. Domestication most likely began with someone separating a few animals from the herds and penning them. Once penned, animals were at hand to be slaughtered on ceremonial occasions or when needed. In a sense, early agriculturalists invented a way to store meat by keeping the creatures alive. Sheep, goats, cattle, and pigs were among the animals they selected. Over millennia, the seasonal rounds of most of the wandering bands grew less and stopped. Gradually settlements began to grow up near and around the domesticated beasts. Increasingly some of the animals were used for milk and wool as well as meat. This led, in turn, to the beginnings of a pastoral agriculture, as gathering of wild fodder plants was necessary to help overwinter animals. The gathering of wild plants was the precursor of their domestication and was the next major important phase in agriculture.

The structure of human societies was totally transformed by plant domestication. Seeds, especially from grains like wheat, rice, millet, and corn, require careful selection. Over a period of ten-thousand years of selecting and growing plants these early people, most probably women, looked for size, taste, and intact seed heads. An intact head was essential so the grain would not be lost when the grass was cut. In a relatively short period of time the intact-headed strains of cereals became largely dependent upon humans for survival. Although no direct evidence for this has been found, it is likely that selected seeds were sown into the ground just after burning to prevent wild plants from outcompeting their domestic relatives. If this was the case, then fire would have been the earliest form of tilling the soil and preparing it for seeding.

Very gradually the growing permanence of penned animals and cultivated crops led to the building of walled settlements and substantial villages. The grains had to be protected from the livestock, and livestock, in turn, from such predators as tigers, cougars, hawks, wolves, and coyotes. The earliest walled architecture developed more likely for agricultural rather than military purposes. Jericho in the Jordan Valley and Catal Huyuk in the central plain of Turkey were two such early settlements, established during the eighth and seventh millennia B.C. respectively. They were situated in the areas where many of the wild cereals originated. A comparable settlement to the east, Tepe Ali Kosh, a village on the steppes at the base of the Zagros Mountains in what is now southwestern Iran, dates back to 7500 B.C. We know that the people there used the seeds of over forty plant species, the most important being domesticated emmer wheat and two-row barley. Analysis of their diets has brought to light the startling discovery that, in terms of nutrition, they ate better than most people today.[2]

These early phases in the development of agriculture modified but did not break down or destroy the great ecological cycles. Domesticated herds did not wander freely like their wild cousins but moved in a proscribed area tended by shepherds. The new seeds required care, but still had to be sown according to the dictates of soil and climate. Burning meadows did not preclude vast untouched regions where climax plant associations like forests dominated the land. The great biological provinces were still primaeval. Human settlements were small pockets in the larger realm of natural life. We cannot imagine what the world was like then when agriculture was becoming a factor in planetary evolution. Our occasional isolated parks and wildlife refuges cannot prepare us for comprehending such a world. It may well be that the people of those days would have perceived subsequent agricultural advances as violating the great natural laws. Some peoples have regarded the use of the plough, for example, as wounding the Earth, ripping open soil and exposing it to the sun, wind, and rain. Yet a simple plough is undramatic and mild in comparison to the great machines we use to farm today.

Around 5000 B.C. in Mesopotamia a step was taken that would alter the pattern of settlements for ages to come. Irrigation began to be practiced in the large alluvial plains of the Tigris and Euphrates River Valleys which were ringed to the north by the Zagros Mountains, and to the north and

west by the Taurus Mountains of Anatolia. The lower reaches of these
mountains were the centers of the origination of wild cereals. In areas re-
ceiving rainfall of about three-hundred millimeters per year, the grains
selected from these cereals flourished. The earliest villages and towns arose
in a narrow band in the foothills of Mesopotamia.

The valleys of the Tigris and Euphrates were too dry for much cultivation of cereals. To the south lay the drier Syrian Desert extending to the
Persian Gulf in the east and to the Red Sea to the south and west. In the
foothills, however, about six-thousand years ago, people began to cut channels from streams across the fields, causing water to spill over onto the land.
The irrigation of crops had begun. It is believed that these early ditches
were dug as a safeguard against the unpredictability of the rains. Gradually
a series of villages whose inhabitants practiced irrigation were built on the
northern rim of the great alluvial plain. With time irrigation methods became more complex. Channels were extended lower into the fertile river
valleys, then beyond, into the semi-arid lands along the course of the Tigris
and Euphrates. These lands slowly became laced with a vast network of
dikes, canals, and dams. Such complex water regimes demanded regulation which, in turn, necessitated organization and regimentation. Centers
which were religious in origin were adapted to regional administrative purposes as well. The resulting high levels of organizational sophistication laid
the conceptual foundations for building public buildings and the increasingly impressive temples. The great Sumerian culture of southern Mesopotamia became established. By 3500 B.C. the Sumerians had invented writing
in order to chronicle their business affairs. Partially as a result of the creation of the vast irrigation complexes, and enough food each season to turn
attention to other organizing principles, civilization was born.[3]

From a historical as well as ecological perspective these methods of
irrigation were a mixed blessing. Large numbers of people were enslaved
to maintain the irrigation systems as channels and impoundments silted up
regularly. In the dry climate, irrigation water evaporated quickly, leaving
salts on the surface of the land. Over time, the productivity of the land
dwindled. Today this region is modern Iraq. Seen from the air, it shines like
a white-washed wasteland, devoid of most life. Much of it was salted beyond
repair long ago by the ancient Sumerians. Between 2000 B.C. and 1000 B.C.
the Sumerian Empire, and subsequently its successor in southern Mesopotamia, the Babylonian Empire collapsed. Irrigation which had made the be-

ginnings possible of cities, literature, and civilization, led eventually to devastation.

The next major advance in agriculture, at least in the modern world, took place in Europe. In the centuries leading up to the first millennium A.D., Europe was wilder and richer in biological terms than it had been in Roman times. Vast forests covered much of the land. Terrorized by Vikings from the north, Saracens from the south, and the Magyars from the west, people frequently were forced to abandon settlements and their small, tilled plots reverted to shrubs and scrub forest. Around 950 A.D. the raids grew less and halted. The population of Europe then has been estimated at about twenty-five million people. It was at about this time that a new kind of plough, the moldboard plough, came into use. Large and heavy, it was pulled by draft animals. Digging deep into the soil, it lifted dirt up and folded it over. The moldboard plough enabled a farmer to till on a previously unprecedented scale—far beyond that of earlier plots which were clustered around settlements and villages. Weeds could be controlled over whole fields. As its use was extended, marshes were drained and forests felled so that more land could be brought into cultivation. In a short span of time large areas of Europe were, in a sense, colonized internally. With only natural rainfall, significant agricultural surpluses were produced. As a result, trade was reactivated and towns, fairs, and markets grew up quickly. In two-hundred years, by 1150, the population of Western Europe had reached about forty million. The wealth generated from the expanded agriculture was sufficient to underwrite the Crusades.

Like every invention, that of the moldboard plough had a shadow side. By expanding the scale of farming it engendered an agricultural working class. Formerly independent villagers began to be reorganized around a landholding system under the authority of a local lord, who was linked through allegiance to a king. This ruler had royal authority over the land. The feudal monarchies, which grew in power, did so at the expense of the peasant. At the time of the Norman conquest of England, free-holding peasants were rapidly being indentured as serfs, many living almost like slaves. Although it is a simplification to attribute so much to the invention of the moldboard plough, it is clear that without it history would have been different. It remained the major tool for tilling land into this century.

By the end of the eighteenth century in England the industrial revolution had spread to the country.[4] Machines were appearing on the farms.

One of the most important was the big steam-driven thrashing machine which greatly expanded the production of grain. In order to maximize agricultural profits English landowners, many of whom were industrialists, wanted to make labor and rents commercially negotiable. As a result, the traditional agricultural workers, the yeomen, were frequently replaced by less skilled, migrant, seasonal laborers. Land ownership was a primary vehicle for change and the landed class supported the technologies that were altering agriculture. The same technologies fueled a crisis in tenureship and created widespread social unrest. By the end of the eighteenth century in England, five-thousand families owned half of the cultivated land, and of these, a nucleus of four-hundred families owned a quarter of the total. A rural population of ten million was dependent on these families and their largess—or lack of it. Increasingly, the new machinery was used to intimidate agricultural workers and to weaken their political position. Perhaps no where is it better described than by Thomas Hardy in *Tess of the D'Urbervilles*:

> Close under the eaves of the stack, and as yet barely visible, was the red tyrant that the women had come to serve—a timber-framed construction, with straps and wheels appertaining—the threshing machine which, whilst it was going, kept up a despotic demand upon the endurance of their muscles and nerves.
>
> A little way off there was another indistinct figure; this one black, with a sustained hiss that spoke of strength very much in reserve. The long chimney running up beside an ashtree, and the warmth which radiated from the spot, explained without the necessity of much daylight that here was the engine which was to act as the primum mobile of this little world. By the engine stood a dark motionless being, a sotty and grimy embodiment of tallness, in a sort of trance, with a heap of coals by his side: it was the engineman. The isolation of his manner and colour lent him the appearance of a creature from Tophet, who had strayed into the pellucid smokelessness of this region of yellow grain and pale soil, with which he had nothing in common, to amaze and to discompose its aborigines.

This demand for speeded up work to feed machines was dehumanizing to the traditional workers. Protests broke out in the early nineteenth century throughout the countryside in the form of the "Bread and Blood" riots and burnings. The slow migration of villagers and country people to the grimy mills and sweat shops of the new industrial cities began. This pattern was not long confined to Great Britain, but spread quickly throughout Western Europe.

As agriculture became more mechanized and more closely tied to commodity markets, it fell into serious economic trouble. By the middle of the century crop failures were added to agricultural debts. In 1800 one-third of the population in England had been working in agriculture. By the end of the century the faction had dropped to one-tenth, representing an agricultural decline which gradually created havoc in Great Britain and Western Europe. The exodus of fifty-million people from Europe, most of whom settled in the United States and Canada, began. The countryside was depopulated in the path of the machine and the land consolidated in the hands of a few men. This process continued until the advent of World War I. This revolution in agriculture of the late eighteenth and nineteenth centuries grew concomitantly with the developing capitalist system. The same people who owned the expanding extracting and manufacturing industries also owned much of the farm land. The values and attitudes towards the natural world they had developed in industry spilled over onto their dealings with the land. Although agricultural productivity increased and the amount of food produced per agricultural worker gained significantly, the system itself was inhumane and therefore flawed. The lot of agricultural workers was not improved by the advent of the new machines.

The most recent and possibly the most powerful of all the innovations in agriculture, the Green Revolution, started in the New World. Although its history is short, its methodologies and principles have spread rapidly around the world. The Green Revolution had its beginnings in the late 1800s in the United States when the federal government established land grant colleges and a countrywide agricultural extension service. Scientific data on agriculture was assembled on an unprecedented scale, tested on the various campuses, and quickly put into practice on farms. Every agricultural region in the country was affected and within a generation or two agriculture was thoroughly modernized. As electricity was introduced in rural areas in the twenties and thirties, it provided an additional catalyst for change, bringing with it more urban lifestyles. The self-image of the farmer changed. Farmers began to see themselves as agricultural businessmen and farming as the production of agricultural commodities. The concept of stewardship, the partnership with and careful tending of the land, faded.

The basic attributes of what in the 1960s came to be called the

"Green Revolution" are well known: the massive use of, and dependency upon petrochemicals, especially fertilizers, pesticides, herbicides, and fungicides; the requirement of a large influx of electricity in many facets of the food cycle; and reliance on the internal combustion engine for all tillage and harvesting. In terms of the scale and diversity of the machinery that came to be used, the Green Revolution is without precedent. Some of the machinery currently employed in farming borders on the bizarre. There is a machine called "Big Bud, Model 747" which costs around half-a-million dollars. It weighs sixty-five tons, pulls a cultivator eighty feet wide at six miles per hours and can plow a thousand acres in twenty-four hours. Big Bud is so large it requires television monitors for the driver to see what is happening with the cultivator out behind.

The true hallmarks of the Green Revolution, however, are the new varieties of plants and breeds of animals. Both plants and animals have been developed which are scarcely capable of fending for themselves without help from an extensive agricultural infrastructure. The trade-off for such dependency lies in the unprecedented yields that are produced per acre. The hybrid corns, short-stem wheats, and new high yielding rices, which have given the Green Revolution its name, have spread throughout the world in the last twenty years. Over one-hundred-and-thirty million acres are under cultivation with these new crops, which require one-hundred-and-twenty to one-hundred-and-eighty pounds of nitrogen fertilizer per acre per year. Two and sometimes three crops are grown a year which, under the best circumstances, has increased productivity two to three times over that achieved by traditional agricultural practices using plot rotation and traditional varieties of seeds. Using huge machines has encouraged specialized crops over large regions like the vast rolling hills of Iowa corn country, or the aquifer-fed center pivot irrigation grain fields of western Kansas.

Modern American agriculture is vital to combating the dollar drain to oil-producing countries through exported surpluses. Science, technology, and banking have joined forces to keep agriculture economically advantageous, in spite of which, it is potentially vulnerable and extremely inflexible. In order to function, it requires, in the right place and at the exact time, a precise combination of seeds, irrigation, electricity, machines, fertilizers, pesticides, fungicides, herbicides, advanced weather reporting, wide range pest analysis, and sophisticated marketing information. It takes a healthy industrial sector in an advanced economy to support this agriculture. If one or two of the key elements are removed, the entire system is threatened.

Most scientists involved in agriculture are not concerned either with changing the structure of agriculture or with questions of land tenureship. Currently, in many areas, the countryside is becoming depopulated. Only three percent of the population are farmers and the numbers of family farms has dropped from 6.8 million at the beginning of World War II to 2.7 million today. Since 1950, every week some two-thousand farmers have been leaving their farms for urban centers. Rural culture has suffered. In parts of the country there is a ghost town feeling to some rural settlements. Because of high capital cost, the new technologies have forced farming to function accordinging to the imperative of the market economy, leaving behind values based on family and land care. This has resulted in exploitation of both land and farm laborers. Furthermore, these methods of tillage and cropping are allowing the land literally to blow away. Current soil losses have been estimated at twenty-five percent higher than during the Dust Bowl years of the 1930s. In the state of Washington twenty pounds of topsoil are lost for every pound of wheat grown. By the year 2000 soil productivity in many areas will be almost non-existent.[5]

In the early sixties Rachael Carson gave warning in her book *The Silent Spring* that we were poisoning the planet with chemical agriculture. Today, as a result of infiltration by agricultural chemicals, many surface and ground-water supplies are unfit to drink. Rachael Carson also predicted that many pests would become poison-resistant and her fears have come true. Three-quarters of the insect pests in California are insecticide resistant.

Farming uses more petroleum, primarily in the form of fertilizers and biocides, than any other industry. Such dependency is tantamount to an addiction. Recently the trend has been for fertilizer use to increase while yields have dropped. Grain yields have started to fall while between 1960-1979 nitrogen fertilizer use nearly quadrupled. Farmers also receive less and less of the food dollar as the number of middlemen–manufacturers, packagers, wholesalers, retailers– has increased. Meanwhile debt structure in agriculture is frightening. In 1980 fifty percent of farm income went to pay interest on farm debt and between 1960 and 1982 farm indebtedness jumped over eight times to a current high of over two-hundred billion dollars. Many farmers are over a million dollars in debt and pay three-hundred-

and-fifty-six dollars a day on money borrowed at a below prime of thirteen percent.

Agriculture may be kept viable in the short term, however, by strong federal price supports. Farmers can be given priority in acquiring petroleum products. New and increasingly toxic pesticides can control the growing pest populations. None of these strategies can be sustained indefinitely, however, in either economic or ecological terms.

2. Agriculture Based on Stewardship

Each of the historical developments in agriculture that we have traced has resulted in an explosion in productivity, with resulting food surpluses. With each major innovation the wild landscape has become increasingly domesticated. Although this continues to be the dominant trend, the beginnings of a very different form of agriculture are discernible—one that is restorative, ecologically inspired, and applicable to the design of future settlements. It is an agriculture based on the values of stewardship and involves both ancient and modern knowledge. Rather than exploding outward as the other innovations have, with biology as the model it will implode, turning inward and moving in the direction of miniaturization. This type of agriculture will break with the past in that the culture of food will be more closely interwoven with the fabric of settlements. As our designs indicate, the division between agriculture and culture will heal as towns, cities, and villages become more agricultural. The utter separation between rural and city life, and the polarization of each, makes each an unsustainable form. The transformation of agriculture, which started as a primarily urban experiment, will soon move outward to reoccupy and care for the land.

The concept of stewardship in agriculture is steeped in different values than the present business perspective. Throughout history people who have worked the land have been serf, peasant, migrant farm worker, yeoman, pioneer, homesteader, tenant farmer, family farmer, and agribusinessman. In each role, social position and self-image affects both behavior and attitude towards the land. Currently migrant farm workers are an exploited class, struggling to earn wages in a market economy. Equally controlled by economic contingencies, the agribusinessman must, of necessity, see labor, land and crops as commodities, and machines and chemicals the primary strategies for maximizing profits. Both for him and for the migrant worker the land has become objectified. Seen through the eyes of the agricultural steward the land and the countryside are living, vital entities held in sacred trust, to be nurtured and protected for the good of all things. He or she lives in the harsh economic realities of the day, but struggles for the delicate balance between ecological necessity and economic prudence. To achieve such a balance such a steward starts on a small scale, applying the precepts that underlie biological design. The steward applies a unique growth principle to agriculture, realizing that the sub-elements of a system cannot be enlarged independently and that in order to grow, the system must duplicate itself so that the internal integrity remains intact and functional. An agricultural steward strives to be a healer of the land. The wounds rent by machines, poisons and bad land practices then become the raw materials of re-creation. The work is to restore waste places.

The social forms of farming can change from very large, isolated farms to smaller holdings. Stewardship in agriculture will require a range of skills and new forms of cooperation. In the past, in this country, most farmers have worked alone with their families usually on their own land. Although this will no doubt continue, other social modes may emerge as well. An ecological farm will be comprised of diverse interacting components derived from horticulture, orchardry, livestock husbandry, entomology, aquaculture, bioshelters, and field crops all linked to create exquisitie three-dimensional landscapes. It is conceivable that such a farm could be owned and worked cooperatively by a team of specialized farmers each of whom runs a subcomponent of the whole. It could have its own economic infrastructure and be set up as a small profit-sharing company. Such an arrangement would allow for a creative balance between individual and collective interests with the advantage of access to shared tools as well as skills.

In Chapter Three we sketched some of the attributes of natural systems and the patterns in which ecosystems organize themselves over time in the process of ecological succession.

Peter Raven in *The Biology of Plants* gives the following description of ecological succession as

> In ecology, the slow, orderly progression of changes in community composition during the development of vegetation in any area from initial colonization to the attainment of a climax typical of a particular geographic area.[6]

In the *Fundamentals of Ecology* Eugene Odum extends the meaning of

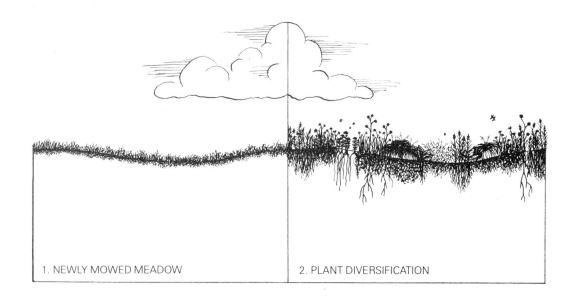

1. NEWLY MOWED MEADOW | 2. PLANT DIVERSIFICATION

**Succession in a Meadow,
Sequence Measured in Decades**

succession to include the implications for human activity and agriculture. He states:

> In a word the 'strategy' of succession as a short term process is basically the same as the 'strategy' of long term evolutionary development of the biosphere, namely increased control of, or homeostasis with, the physical environment in the sense of achieving maximum protection from its perturbations. The development of ecosystems has many parallels in the developmental biology of organisms and also in the development of human society.[7]

Studies of ecosystem strategies like those of Dr. Odum create design blueprints for stewardship agriculture. Such blueprints permit agriculture to move toward becoming intensified, miniaturized, and diversified, fostering a dimension of self repair and sustainability. Market gardening, aquaculture, tree crops, and livestock are important components. Each of these supports and enhances the production of the whole. Like the farm on Java we described, multi-story diversification, close planting, and mutually re-

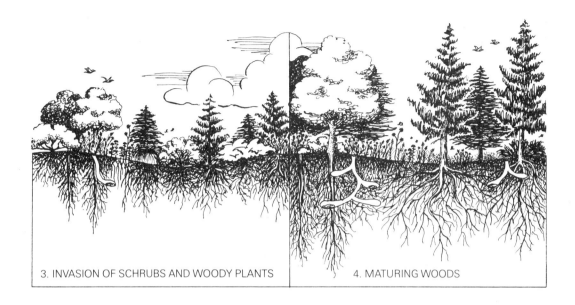

3. INVASION OF SCHRUBS AND WOODY PLANTS   4. MATURING WOODS

inforcing intercropping are integral elements. Light, nutrients, wastes, moisture, and beneficial organisms like bees, are shared by many of the components.

Successional or ecological agriculture differs from ordinary farming in that it adapts to changes over time. In early phases, annual crops and fish ponds might dominate the landscape, but as the landscape grows and matures, a third dimension develops as tree crops and livestock come into their own. The key is to mirror the natural tendency of succession which, over time, creates ecosystems that are effective and stable utilizers of space, energy, and biotic elements.

A successional farm is most likely to be a small acreage farm intensively worked. The opposite of monoculture, it uses a number of diverse elements to establish the symbiotic relationships which lead to overall system productivity, health, and integrity. Components will include marketable fish, vegetables, herbs, forage crops, bees, poultry, livestock, fruit, nuts, and other economic trees. They will be assembled so that as the farm evolves through each state, it will become an increasingly viable and cost effective system. Each year fewer and fewer inputs will need to be imported for maintenance and self-regulation. Systems dynamics has given the agricultural steward the ability to model time, space, economic pathways, and ecological needs. At New Alchemy we studied computer models derived from wild ecosystems, then introduced domestic elements. This approach makes it possible for the landscape to retain a deep ecological integrity even as it becomes increasingly agricultural.

Reflecting the natural world, ecological agriculture is complex—the building blocks or subcomponents are comprehensible. To begin requires gardening skills, a basic understanding of home fish aquarium techniques, some knowledge of orchardry and small animal husbandry, and an elementary ability to identify insects. It is, in some aspects, far more accessible than its Green Revolution counterpart. The secret is biological information supported by an ecologically derived infrastructure that gradually replaces machinery, capital, space, and large energy and control inputs. The new agriculture is made up of the following seven elements:

1. Bio-intensive Soil Management. One of the most successful methods of soil cultivation appropriate to new forms of agriculture was developed by the biodynamic gardener Alan Chadwick. He brought to North America the raised-bed techniques used in the late 1800s in gardens around Paris and combined them with composting and the somewhat esoteric biodynamic methods developed by Rudolph Steiner and his associates in the 1930s. Chadwick was a teacher. His work has been extended and made accessible to large numbers of people by former systems analyst John Jeavons, whose studies at Ecology Action in Palo Alto, California, are little short of revolutionary. In his book *How to Grow More Vegetables*, Mr. Jeavons advocates cultivating the soil to a depth of twenty-four inches by double digging with either a shovel, or the the deep soil cultivator which he invented. The surface soil is not turned under but set aside while the lower soil is being dug. The surface soil is then replaced and thus remains on top, untrammelled, which allows it to breathe and lets the rain, with its gases and nutrients, percolate down slowly rather than wash surface particles away.

2. Intensive Planting Techniques. Bio-intensive gardening employs a close planting pattern of crops, and often, crops grown in combination. It incorporates abundant use of natural compost, light daily waterings, and avoidance of soil compaction. Under this regime, aerobic microbiotic life predominates in the soil and, by feeding the soil in the right combinations, nutrients are made available for the crops by the microbes. With these soil management and planting techniques, Jeavons has proved that, in California, a complete, nutritionally balanced diet can be grown in a four-month growing season on as small an area as 2,800 square feet per person. Intensive Japanese farming demands twice as much space, American mechanized agriculture requires five times the land area, and low input Indian agriculture fifteen. Jeavons' research has shown that in a six-month growing season a backyard gardener can grow a year's supply of vegetables and soft fruits (three-hundred twenty-two pounds) on one-hundred square feet working on an average of five to ten minutes a day.

Jeavons' studies have given micro-farming an economic dimension as well. A tenth-of-an-acre farm in California conceivably could net ten- to twenty-thousand dollars annually with the farmer working forty hours a week for eight months of the year. One of the reasons for such economic viability is the lack of machinery and non-biological inputs. Design and handwork are substituted for hardware. Jeavons hopes to prove that as much food can be grown on a per hour basis by hand cultivation as is now produced by mechanized commercial agriculture.

The research of Robert Rodale and associates who publish *Organic Garden and Farming* and *New Farm* magazines, make a compelling case for ecological agriculture. For many years Mr. Rodale and his father, J. I. Rodale were among the few advocates of agricultural alternatives in North America. Guided by Robert Rodale's vision of a regenerative agriculture, the Rodale Research Center has experimented with composting, soil building, and water conservation on both garden and farm scales. They have documented their work and chronicled the economic potential of ecological concepts in agriculture.

3. Aquaculture. The solar aquaculture based on the translucent cylindrical solar-algae ponds, which we developed at New Alchemy, is the aquatic analog of bio-intensive agriculture. The three-dimensional, above ground ponds receive light from all sides and the top. The results of the experiments with these tanks have been one of the most productive forms of low energy aquaculture in the world. Fish, like tilapia, obtain a good part of their diet by feeding upon the algae. A five-foot high by five-foot in diameter tank produces up to sixty pounds of fish a year and we expect to almost double this figure one day by integrating new species of fish. In the summer half-grown, the tank-raised fish are transferred to shallow outdoor ponds for fattening to market size. The fish can be rotated among ponds and the pond bottoms, having been enriched by the fish, can be used occasionally as weed-free vegetable gardens. The water from both solar-algae and outdoor ponds is used to irrigate and fertilize agricultural crops. When solar tanks are placed in sunny locations inside bioshelters they serve double duty by trapping and storing solar heat to function as low temperature "furnaces." They are effective enough to provide the heating for New Alchemy's bioshelters when the sun is not shining. Aquaculture can provide part of the early economic base for a farmstead while it is being planted to slow maturing crops like fruits and nuts. Although fish farming has been long neglected in the Western world, it has been a great ecological stabilizer in parts of the Orient and has enormous potential for ecological agriculture here.

When there are ponds or lakes on the land, cage culture aquaculture is worth considering and is easy to integrate with the agricultural landscape. The fish are kept in small floating cages in the pond or lake. The beauty of cage culture lies in the fact that while it scarcely impinges upon the environment, it is productive and it does not demand a high degree of expertise. The pond does much of the caring for the fish, providing habitat and some food. The body of water can still be used for recreation including fishing for non-caged fish and for the irrigation of agricultural crops.

At New Alchemy Bill McLarney and Jeffrey Parkin experimented with cage design in a small pond that borders the Institute and successfully raised good harvests of bullhead catfish and sunfishes in floating cages. They summarize their experiences, as well as those of other researchers, in *The New Alchemy Backyard Fish Farm Book.*[9]

4. Bioshelters. New Alchemy's bioshelters and the bioshelters illustrated in Chapter Four are the solar age equivalent of the nineteenth century barn. Then the barn allowed the farm to continue to function after the first hard frost in fall until the last frost of spring by storing the biological material that sustained the farm. Grains, hay, and silage were kept to feed animals, and livestock were sheltered. In this way the period between growing seasons was weathered by overwintering. The bioshelter does the same thing, only more dynamically. The growing season continues inside the structure. Vegetables and fruit are grown and harvested, fish hatched and maintained, and trees propagated. The bioshelter is the epicenter of an ecological farm. Seedlings, fish, beneficial insets, and young trees can be grown for subsequent outdoor culture. A short growing season is less of a handicap to raising food. Although many people, on first viewing a bioshelter, see it as a technological implant, set apart from the agricultural landscape, we envision it being used increasingly as a source of biological material for the whole farm as well as the most efficient means of bridging the gap between growing seasons.

5. Small Plot Grains. It is popularly held that grain culture belongs only in the vast fields of the midwest and west. At the Rodale Research Center Robert Rodale and Richard Harwood have been promoting backyard scale grain raising for years, and in the process have started to change this image. John Jeavons, by growing wheat and rye in one-hundred square foot plots using bio-intensive methods, is drawing even more attention to new options in backyard farming. His first experiments have yielded twelve pounds of grain per one-hundred square feet. In California they expect to be able to obtain two, twenty-six pound crops over an eight-month season—which is equivalent to fifty-two loaves of bread from a ten foot by ten foot plot. Mr. Jeavons threshes the grain with small threshers are built in Japan.

At the Land Institute in Kansas, Wes and Dana Jackson and their associates are experimenting with a more radical concept for grain, one which could transform the way in which the prairies are farmed. They are trying to develop perennial grains which would yield crops like domesticated annual wheat and rye. These long-lived perennial grasses would protect and enhance the soil, as the wild prairie grasses do, but still produce an annual crop of edible grains. One of their new plants is four plants in one. The chromosomes come from eastern gamma grass, a wild relative of corn, domestic corn, and a perennial wild corn. They have been combined to create one plant, in the language of genetics called a haplotriploid. The Jacksons are aiming for the hardiness of eastern gamma grass, the yielding ability of domestic corn, and the perennial nature of the wild Mexican corn and eastern gamma grass combined.

The Jacksons have also developed a hybrid between sorghum, a perennial which they intend to make winter-hardy by breeding it with sugar cane. The high sugar content sorghum/sugar has an "anti-freeze"-like substance in its tissues, which increases its hardiness. They have been back crossing it to the sorghum and breeding this plant in turn to Johnson grass, a tough plant with a deep penetrating rhizome. The result is a winter-hardy sorghum grain with a deep, penetrating root system. In another successful experiment the Jacksons searched through six species of high yielding perennials from the world's collection of 4,300 varieties and found a plant from the shores of the Soviet Union at about the latitude of Anchorage, Alaska. It was a wild perennial rye, known as the giant wild rye *Elymus giganteus*. It is known to have been eaten by the Mongols. They crossed this giant rye with wheat. The cross has survived two Kansas summers, one of which brought the worst drought since the Dust Bowl of the thirties.

Another of their projects involves changing the sex ratio of eastern gamma grass so that it becomes mostly female and therefore highly yielding. The wild easten gamma grass, a perennial relative of corn, has grains which contain twenty-seven percent protein, three times that of corn. It has almost twice the thiamine of corn. The drawback is that, like most perennials, it is a low yielder. The reason for the low yield is the inflorescence or flower cluster which is mainly made up of tiny males. An astute botanist discovered a mutant in the midwest that was mostly female. The Land Institute acquired a clone, which, in tests, has yielded fifteen to twenty-five times that of the normal gamma grass. This is equivalent to thirty bushels an acre—which rivals the yielding capacity of wheat. With sufficient high yielding perennial grains, a prairie agriculture could emerge which would make the plough a thing of the past. Wildness and human intervention would once again be more closely balanced.[10]

6. Livestock and Poultry. Animals are integral to any natural system. They are nourished by plant and microbial life and contribute to the cycle with their excretions and eventually with their decomposing bodies. Modern agriculture removes animals from large tracts of land to leave the fields free for monocrop soybeans, corn, peanuts, grains, or hay. Although there are still some animal pastures, increasingly livestock and poultry are confined to feedlots or chicken factories. Although many ecologically sensitive farmers are vegetarians and interested only in plants, animals are as much a part of the ecological fabrics as plants or microbes, and a healthy agricultural landscape depends upon them. Successional farming requires a balance between plants and animals, but the movement of animals must be carefully regulated. In early successional stages on an agricultural landscape chickens, ducks, and geese should predominate, as their feeding controls weeds and the manure fertilizes the soil. Whereas in urban areas, poultry may be the only animals it is possible to raise, on larger rural farms, when the trees mature, pigs, sheep, and cattle will be part of later successional stages. Particularly in cold climates where growing seasons are short, milk, eggs, and meat have long been the backbone of human diets. In a sustainable agriculture they may not be as predominant with more fish, vegetables, and fruits being produced, but they will retain their niche in the overall ecological integrity of the farm. One of the challenges of the future will be to breed animals adapted to agricultural ecosystems, while still retaining some of the improvements of animals breeding in this century.

7. Agricultural Forestry. In areas where it is appropriate to the bioregion, the climax state of stewardship agriculture is farming in the image of the forest. The prairie, desert, and savannah similarly each dictate their form of sustainable agriculture. Where a landscape is mostly wooded, the agricultural element will include fish ponds, annual vegetables and herbs, perennial grains and herbs, soft fruits, livestock and bioshelters as well as trees. When the fruit, nut, and fodder trees are mature and become the predominant element, they will provide much of the economic base for the farm.

Over fifty years ago the American geographer J. Russell Smith studied the relationships between trees and human communities and found a direct correlation between planting and caring for trees, and healthy diverse communities. His research spanned several continents and many cultures. He chronicled the negative effects of deforestation on human beings. In his best known book, *Tree Crops*, he advocated an emphasis on trees as the foundation for a restorative agriculture.[11]

More recently, Bill Mollison, an Australian who comes from Tasmania, has been teaching and popularizing a system of farming based on tree crops which he calls Permaculture. His books *Permaculture One* and *Permaculture Two* go beyond the concepts of J. Russell Smith in that his perennial agriculture includes a broad ecological infrastructure set in a forested landscape. He stresses the importance of water, grasses, insects, wildlife, livestock, species relationships, and siting, in the overall planning of an agriculatural landscape.

In *Permaculture Two* Mollison states:

> I regard permanent agriculture as a valid, safe, and sustainable, complete energy system. Permaculture as defined here claims to be designed agriculture, so that the species composition, array and organization of plants and animals are the central factor.[12]

By insisting on dealing with complex biological systems as integrated wholes, Mollison flies in the face of those who would keep information in narrow specializations. He claims that there are enough agricultural elements to fill all the necessary roles in an ecosystem. At New Alchemy under John Quinney we have created a computer model for a theoretical farm in both economic and ecological terms. It is projected over a twenty to thirty year time frame. We are trying to locate biological patterns that assist ecological cycles and optimize the overall health and productivity of the maturing farm model. The grafting of bio-intensive methodologies with orchard crops represents a new direction for tree farming. So far, even in these early experimental stages, we are seeing exceptional tree health and productivity. Two dwarf apple trees on eight-foot centers, for example, have yielded two-hundred pounds of fruit. In statistical terms the average person eats one-hundred-sixty-two pounds of tree fruit annually. Using bio-intensive methods, a twenty-foot by twenty-foot plot could feed a family the fruit that it needs.

In getting started with the process of restoration, ecological agriculture will probably have to continue to use some fertilizers and heavy machinery, particularly on worn-out land. Machinery for chiselling, terracing, pond digging, mowing, and contour plowing may be necessary. In the early phases land will be farmed less intensively, as it will need time to heal. On medium-sized farms, especially in hilly areas, oxen and horses could be used extensively as draft animals. Ecological agriculture will need to turn to composted city and town wastes to build soil fertility. It will introduce atmospheric nitrogen into the soil by growing nitrogen-fixing legumes like beans, peas, and vetch. In some instances, the farmer may revive the use of fire to cold burn in early spring or at the beginning of a rainy season to recycle nutrients or clear overgrown areas. In some parts of the country, fire may be a substitute for plowing, with seeding following the fire. Wes Jackson has experimented successfully with this at the Land Institute. We burn our valley every spring.

Extensive research into beneficial plants and organisms is needed for integration into the larger ecological framework. Weeds must be studied, as they are reliable indicators of soil health and guides to crop planning. The technological hallmark of ecological agriculture will be the incorporation of renewable energies in which sun, wind, and biofuels will come to play a significant role. Gradually energy needs should drop to a point where renewable sources can readily meet all demands. Electricity use, for example, can be reduced from heavy demands, like deep well irrigation, to a more modest role in powering control devices, electric fencing, and lighting. The solar cell will continue to find application, as it is ideal for remote site day-time electricity production.

8. Wildness and Technology. Further ingenious innovations will continue to surface. Ocean Arks' Joe Seale has designed a highly efficient wind-powered refrigeration and ice-making machine. The machine, designed initially for fishing communities in the tropics, will produce ice, cold storage, and auxiliary heat. The development of machines like this are reassuring because, of all energy sources, only renewable forms are distributed equitably around the globe. People everywhere have access to some combination of sun, wind, or rain, and the technology using these is advancing rapidly. There is easily enough energy to go around and to rebuild agriculture in an equitable and humane way.

Working within this framework of an agricultural ethic that is resolved to do much with little, the study of weather near the ground, microclimatology, looms large in importance. An ecologically-minded farmer sees each patch of ground as unique, mysterious, and capable of a full flowering at the hand of the steward-designer. Light, shadow, moisture, drainage, and air flow must all be incorporated into agricultural design. At the same time, wildness and wilderness must be revered and tentacles of wildness be allowed to permeate, like the threads of a tapestry, throughout the agricultural landscape extending even into the heart of cities. An ultimate goal might be that for every acre which is farmed another would be set free. If we were to create wilderness belts winding like highways throughout the country in continuous bands, it would be possible to walk from Cape Cod to California, and from the Gulf of Mexico to Hudson Bay by following these belts. Such areas would help to protect endangered species and reintroduce the wild side of nature into culture, recalling Gary Synder's image of "computer technicians who run the plant part of the year and walk along with the elk in their migrations during the rest."[13] By making the wilderness accessible in all parts of the country, the indigenous qualities of each bioregion will become increasingly apparent. As a start, New Alchemy bioshelter agriculturist Colleen Armstrong has experimented with what we call ecological "islands," clusters of plants and beneficial animals left untouched and unmanaged inside bioshelters. Control agents for pests like lady beetles and little lizards take shelter in the "islands" where they are protected from regular tilling and harvesting in adjacent agricultural zones. Such predators range out in search of crop pests but can return to the safety of the stable ecosystems of the island. A hedgerow serves a comparable function outdoors on a farm. On a larger scale, fenced ecological "islands" in patchwork patterns to protect the wild grasses, flowers, birds, and beneficial insects characteristic of more stable landscapes, and to provide a continuous supply of genetic material, are an important element in co-evolutionary restorative agricultural strategies.

The computer is a form of technology well suited to the ecological steward. Because natural processes often are not obvious but counter-intuitive, modelling enables the designer to store and retrieve large amounts of information. Frequently in our work computer models have given us access to the "memory" of the living system we have monitored and tended. We can ask such questions as what is the optimal ratio of pond area to pasture or to orchard trees. How many fish and what combinations of species are optimal for our summer climate? The evolving model becomes a book which an agriculturalist can read and redefine. Although the real work is on the land with living things, there is a place for a small computer, loaded with the best ecological and agricultural information in working with models. Such models are meaningful, however, only as they become ever closer to nature.

Evidence in support of the viability of ecological agriculture is beginning to accumulate. Paul Hawken in *The Next Economy* predicted a trend back to smaller farms and an agriculture based on information and human skill as people grasp that there are forms more sound economically than large scale, capital and oil-dependent mechanized agriculture.[14] A study by the Institute for Environmental Studies at the University of Wisconsin found that a landscape made up of communities of around thirty-five thousand people surrounded by bands of intensively cultivated land is most suited to a society based on renewable energies.[15] Such studies reinforce both the economic and ecological good sense of a return to an agriculture based on stewardship to accompany the redesign of urban and settled areas.

Chapter Six

The Transforming Energy

It was our hope, in putting together this book, to make a persuasive, documented case for the ecological design of human settlements and the agricultural landscapes that support them—one which would enhance the possibilities for a sustainable, long-lived future. We have offered a sampling of the range of the fledgeling undertakings now becoming visible. These are still only pinpoints on a vast map but shape patterns that could help us chart our way through the unknown in the coming years. Tentative and isolated as they are, each has key elements in helping us work out the restructuring of how we live. They are signs that are hopeful and stand out in a period like this, which might, given the most optimistic analysis, be pronounced a time of transition. That we are in such a period is indicated not only by the pace of change itself but by the number of intellectual and informational examinations of our fate. We think of the 1974 study *The Limits to Growth* by Dennis and Donella Meadows,[1] the *Global Report on the Year 2000* prepared for President Carter by the Council on Environmental Quality,[2] the book *Seven Tomorrows* by Paul Hawken, James Ogilvy and Peter Schwartz based on data compiled by Stanford Research Institute,[3] and a proliferation of other future studies. There is also a tendency now in popular culture to examine our past as well as our future, as do the book *Clan of the Cave Bear*,[4] and the film *Quest for Fire* which have attracted large numbers of readers and viewers. The films *War Games*, *Testament*, and *The Day After* take an unflinching look at the present nuclear danger. The popularity of space age science fiction, although frequently escapist, attests to a widespread realization of a changing future. It almost seems as though we are experiencing a widespread, collective identity crisis, searching back and forth in time for clues as to where we may have come from and where we may be bound. So much that is familiar seems to be slipping our from underneath us. Very little remains constant for us to cling to. In our personal lives, the solidity of the family, sex roles, values are in upheaval, and many peoples' lives seem to be in a state of on and off crisis. It is Theodore Roszak's theory that the emerging need of the person is the Earth's urgent cry for rescue. In his book *Person/Planet*, he writes:

> What, then, does the Earth do? She begins to speak to something in us—an ideal of life, a sense of identity—that has until now been harbored within only an eccentric and marginal few. She digs deep into our unexplored nature to draw forth a passion for self-knowledge and personal recognition that has laid slumbering in us like an unfertilized seed. And so, quite suddenly, in the very heartland of urban-industrial society, a generation appears that instinctively yearns for a quality of life wholly incompatible with the giganticism of our economic and technological structures. And the cry of personal pain which that generation utters is the planet's own cry for rescue, her protest against the bigness of things becoming one with ours. So we begin to look for alternatives to that person-and-plant-crushing colossalism. We search for ways to disintegrate the bigness—to disintegrate it *creatively* into humanly scaled, organically balanced communities and systems that free us from the deadly industrial compulsion of the past.

In the public arena there is widespread loss of confidence and disillusionment with the world situation as created by competing nation-states and with the ineffectuality of governments to ease the ongoing sense of crisis. That we are living through a time of transition, unstructured and uncertain, seems unarguable. The question most pertinent for us to ruminate on becomes—transition toward what?

There are a good many scenarios about the future to choose from these days. Very few of them are cheerful. The one implicit in the operating of almost all governments is that through faith in business as usual and in technological advancement, we will muddle through this period without fundamental change and that things gradually will begin to get better. The worst, equally implicit to governmental policy, is a reckless gamble against extinction by either nuclear holocaust or ecological disaster. There is, between the two, a sort of mid-way possibility—a slow guttering out of industrialism to be followed, in the dim future, by some now unforeseen remnants of our culture. Yet another is that we would be seen to be clearing the stage for cultures that have remained in the wings through this most recent act in the human drama.

Preoccupation with the future, forecasting and advocating decisions that will be binding beyond one's own time is a tricky business. Hubris is not the least of the pitfalls. The only safe guideline, as far as we can see, is that, whatever our decisions, they attempt to open options for ourselves as a species and for the diversity of the other species with whom, ultimately, we have a covenant and with most of whom, we share a common fate. As the Gaia hypothesis delineates very clearly, life on Earth has been and always will be a process of coevolution among all life forms. Both practical and ethical considerations argue that we continue to remember this as we try to find our way in the uncertainty which in the coming years, whatever the scenario, inevitably lies ahead.

Dr. James Lovelock, co-author of the Gaia hypothesis, posed the question: "To what extent is our collective intelligence also a part of Gaia? Do we as a species constitute a Gaian nervous system and a brain which can consciously anticipate environmental changes?"[6] The sophistication of our technology unquestionably is beginning to put us in a position, at least partially, to do so. Information technologies currently used for spying and for perfecting the deadly accuracy of modern weapons have resulted in what Daniel Deudney[7] of the Worldwatch Institute has called a "rudimentary planetary nervous system...a literal wiring of the earth." The same sensing, communication, and data processing technologies, however, could be used with equal efficacy to guard against breaches in international disarmament agreements. Expanding vastly on the work we have done in modeling the bioshelters we have worked with, just as relevant an application of this type of technology as Dr. Lovelock suggested, is the monitoring of the biosphere, not only for military purposes but as watchful stewards of the planet's well-being.

The largest qualifier on possibilities for the human future has only recently begun to be widely talked about or 'named.' Almost without exception, all the books, studies, and theories issued on the subject of the future early on contain a phrase to the effect of "short of nuclear war," or "barring a nuclear holocaust" as a proviso to the other eventualities to come under discussion. Steve Baer put the situation as succinctly as anyone has in *CoEvolution Quarterly* when he wrote:

> A dog that hasn't been chained up long forgets. It rushes across the yard and then–bang. Today when people become excited about the future and involve themselves with new uses of technology they often get carried away with hope–then bam–they think about the bomb, the H bomb, the ballistic missile. Today, like the dog, we all have the chain on

us. There is nothing very marvelous going on unless it is something to untie that chain.[8]

Unthinkable has been the adjective most frequently prefaced to any discussion of the possibility of nuclear war. It has been unthinkable because actually to think about the death of everyone we love and our planet devastated is unbearable. Yet if we are to survive, we must face the unbearable in order that we may not have to live with it. Most people know this at some level of their consciousness, but until recently most of us have successfully screened it out. Since the dawn of the nuclear age we have lived as though comatose in the face of such a reality. It has been less painful to bury the fear in an absorption with the demands of everyday living, thereby laying the ground for a future that daily becomes less likely. We have done this on the level of the individual and on collective and national levels. But even as we have maintained an ostrich-like stance of avoidance, the threat has not gone away but has grown greater daily. To ignore the escalating likelihood of a nuclear confrontation as weapons are steadily stockpiled is to guarantee the highest odds that it will happen, if not by intent, then by human or computer miscalculation, or communication failure. It is uncomfortable to speculate on how long our luck will hold, for once the first bomb has been detonated, it is impossible either to predict or to control what will follow, what chain reaction will occur, or where it all might stop. Military and government strategists talk of containment and of a tactical strike, but there is no one, anywhere, of any type of expertise, who can maintain with any credibility that, once launched, however it might happen, a nuclear exchange could be contained.

In his book *The Fate of the Earth*, Jonathan Schell wrote:

> Bearing in mind that the possible consequences of the detonation of thousands of megatons of nuclear explosives include the blinding of insects, birds and beasts all over the world; the extinction of many ocean species, among them some at the base of the food chain, the temporary or permanent alteration of the climate of the globe, with the outside chance of 'dramatic' and 'major' alterations in the structure of the atmosphere; the pollution of the whole ecosphere with oxides of nitrogen; the incapacitation in ten minutes of unprotected people who go out into the sunlight; a significant decrease in photosynthesis in plants around the world; the scalding and killing of many crops; the increase in rates of cancer and mutations around the world, but especially in the

targeted zones, and the attendant risk of global epidemics; the possible poisoning of all vertebrates by sharply increased levels of Vitamin D in their skin as a result of increased ultraviolet light; and the outright slaughter on all targeted continents of most human beings and living things by the initial nuclear radiation, the fireballs, the thermal pulses, the blast waves, the mass fires, and the fallout from the explosions; and, considering that these consequences will all interact with one another in unguessable ways and, furthermore, are in all likelihood an incomplete list, which will be added to as our knowledge of the Earth increases, one must conclude that a full-scale nuclear holocaust could lead to the extinction of mankind.[9]

Over our heads constantly then, hangs not only the threat of our own deaths but responsibility for damage to life on a far broader scale than humanity has inflicted ever before. And with the eventuality of human extinction, snuffing out the life of all present generations and those yet unborn would come an abrupt conclusion to Gaia's co-evolutionary experiment with human consciousness.

It would be unimaginably wonderful to wake one morning to learn that, like a bad dream banished by the light, the nuclear threat had disappeared, and that we were free from being dogged by the feeling, however deeply buried that, as one young friend of ours put it, "It hasn't happened yet." We could begin to think and feel and fully inhabit the world and time again. But this cannot happen so readily. We have lost our innocence. Foreseen in the meta-reality of myth which travels backward and forward in time, as in the metaphor of the Garden of Eden, we have eaten the fruit of knowledge of good and evil and having done so, nothing can ever be the same again. We are faced with the ultimate, the consequences of our act. The same discoveries that have freed us from the conceptual limitations of Newtonian physics and given us the knowledge of how to free the energy contained in mass have faced us with our mortality as a species, and have changed forever our relationship to each other and to the Earth. Lewis Mumford observed in *The Pentagon of Power*:

> Descartes forgot that before as he uttered these words 'I think'...he needed the cooperation of countless fellow beings, extending back to his own knowledge as far as the thousands of years that Biblical history recorded. Beyond that, we know now that he needed the aim of an even remoter past that mankind too long remained ignorant of: the millions

of years required to transform his dumb animal ancestors into conscious human beings.[10]

The rationale for having landed ourselves in such an incredible dilemma, ironically, is survival. We have not yet thoroughly integrated into our thinking the fact that the use of war as the ultimate political strategy became obsolete when the atom made it impossible to win at war. A pre-Hiroshima mentality remains fossilized in the minds of military planners, strategists, weapons contractors and manufacturers, vestigial from a period that extended from the first raids and clashes between early trading bands to the wars of great empires and nation states. Now the fate of most or all cultures lies with men who personify the extreme institutionalization of masculine values and whose attitudes reflect what George Kennan, the former American Ambassador to Russia, called an "almost exclusive militerization of thinking." This, he says, is symptomatic of what he further terms an "intellectual primitivism and naiveté" unpardonable in a great government."[11] In the January 21, 1982, *New York Review of Books* he wrote:

> And we shall not be able to turn these things around as they should be, on the plane of military and nuclear rivalry, until we learn to correct these childish distortions—until we correct our tendency to see in the Soviet Union only a mirror in which we look for the reflection of our own virtue—until we consent, in all its complexity and variety, embracing the good with the bad—a people whose life and aspirations, whose successes and failures are the products, just as ours are the products, not of any inherent inequity but of the relentless discipline of history, tradition, and national experience. Above all, we must learn to see the behavior of the leadership of that country as partly the reflection of our own treatment of it. If we insist on demonizing these Soviet leaders—on viewing them as total and incorrigible enemies, consumed only with their fear of hatred of us and dedicated to nothing other than our destruction—then, in the end, that is the way we shall assuredly have them—if for no other reason than that our view of them allows for nothing else—either for them or for us.

Leaders of the West seem subject to comparable projection on the part of the Russians. As the Jungian psychotherapist, Dr. Marie-Louise von Franz observed:

> The contemporary division of society into 'right' and a 'left' wing is nothing but a neurotic dissociation, reflecting on the world stage what is happening to the individual modern man: a division within himself,

which causes the shadow–that is, what is unacceptable to consciousness–to be projected onto an opponent, while he identifies with the
abstract picture of the world offered by scientific rationalism, which
leads to a constantly greater loss of instinct and especially to the loss of
'caritas', the love of one's neighbor so sorely needed in the contemporary world.[12]

Like isolated survivalist groups, like the fearful little creature in Kafka's
short story *"The Burrow,"*[13] who closes himself off from all options, anxiety
has led to a mutual blind, self-defeating paranoia that denies the realities
implicit to the larger situation, the outcome of a nuclear exchange. As
Jonathan Schell says:

> ...The upholders of the status quo defend the anachronistic struc-
> ture of their thinking and seek to block the revolution in thought and in
> action which is necessary if mankind is to go on living.[14]

It may seem absurd, even simplistic to say, in such a dire and complex situation, that there are, basically, two choices–in the same sense as the
old fellow meant two when he said he knew two tunes–one was "God Save
the Queen" and the other wasn't. Of the choices that confront us, one is a
nuclear exchange with the near certainty of death for all the civilized world.
To do nothing and to allow the arms race to continue unabated while
weapons technologies become ever more advanced and deadly and move
out into space is tantamount to choosing this. To try to defuse the arms race
with efforts, however frail and quixotic seeming, is to make a holocaust less
of a certainty. Should we lose the nuclear gamble we do not have two
worlds, one for experiment and the other for control. To work to dismantle
the weapons and to forge a world order in which they will not be used, a
hitherto unlikely seeming eventuality that is begining to touch hearts and
minds everywhere, is the only choice that does not make a travesty of
human destiny.

There is taking place at the time of this writing a universal, almost
miraculous awakening to the imminence of the nuclear risk and a fierce
grass roots resolution to oppose it. Growing numbers of Europeans are resolved that their countries will not be the battleground for a Soviet-Ameri-
can exchange. In this country the demand for a bilateral freeze on nuclear
weapons has swept across the country, reaching the floor of Congress in
just over two years from the time it was initiated. The freeze petition caused
something of a stir in our own town of Falmouth. Some people maintained
it was not a local issue. However, an editorial in our local Falmouth news-

paper, the *Enterprise*, pointed out succinctly, "But it is."[15] The townspeople
of Falmouth later voiced their agreement with that editorial statement at
town meeting when the demand for a nuclear freeze received strong backing. The moderator of the meeting was instructed to "notify President
Reagan and the Congressional delegation asking for a moratorium with
the Soviet Union and that all funds for nuclear weapons be transferred to
use for betterment of towns, villages and cities."[16] As one of the leading proponents of the freeze locally, Amelie Scheltema put it, "The charge of town
meeting is the health, safety, and welfare of the town of Falmouth. The federal arms program directly affects all three. Given a choice, let us choose
life."[17] The voters of Falmouth subsequently echoed that sentiment with a
four-to-one margin in favor of the moratorium.

Margaret Mead spoke often of the need for symbols, images, and
common interests compelling enough to transcend national and cultural
barriers. Dr. Helen Caldecott has rallied concerned doctors to revitalize
Physicians for Social Responsiblity and many groups of professionals have
formed similar organizations, such as Educators for Social Responsiblity. A
second anti-nuclear medical organization, Physicians for the Prevention of
Nuclear War has since been formed. Groups form almost monthly—Parent-
ing In A Nuclear Age, various freeze and disarmament groups. We should
attempt to unite parents and poets, painters and plumbers, for a nuclear
free world. The churches are becoming increasingly active and involved. It
was, in part, public pressure that forced the gradual wind-down of the war
in Viet Nam. Comparable pressure put an end to the atmospheric testing of
atomic weapons in the United States and Russia. This is precedence for a
popular voice achieving its objective. The strong grass roots base of the demand for a nuclear freeze as an initial step toward universal disarmament
seems the most promising development of the early part of this decade to
stop an otherwise inevitable war that no one anywhere wants.

In the late sixties, as the ecology movement was beginning to gain
momentum, the poet Gary Snyder, drew up a manifesto he called "Four
Changes". The areas in which he advocated fundamental change were
those of population, pollution, and consumption. These would lead eventually to the fourth of the changes—transformation. He wrote:

> Since it doesn't seem practical or even desirable to think that direct
> body force will achieve much, it would be best to consider this a continu-
> ing 'revolution of consciousness' which will be won not by guns but by
> seizing the key images, myths, archetypes, eschatologies, and ecsta-
> cies so that life won't seem worth living unless one's on the transforming

energy's side. We must take over 'science and technology' and release its possibilities and powers in the service of this planet—which, after all produced us and it.[18]

For our own part, from the earliest days of New Alchemy to the launching of our most recent ventures in planetary healing through The Company of Stewards, and small, sail-powered cargo and fishing boats through Ocean Arks International, there has been much, if not all of the time, an underlying sense of shared excitement and involvement, of being a part of and gaining strength from transforming energy. Our field of endeavor has been in trying to work out the biological and technological infrastructure underpinnings for an enduring society. New Alchemy's role has been to conduct the research that corroborated our intuition of the enormous lying in a growing partnership with the natural world. In order to further safeguard the coming years there are a large number of other developments that must take place. Lester Brown in *Building A Sustainable Society* notes that we must stabilize the world's populations, conserve and preserve resources to a degree we have not imagined, maximize conservation and preservation of resources, change our use and spending of the global energy budget and build sustainable transportation systems. We must, he says, reverse the trend toward urbanization and revitalize agriculture. We must look to greater local self-reliance of towns and cities and simpler life styles for the affluent.[19] The combined momentum of all these changes, he contends, would facilitate slowly the transfer from a growth dependent economy to a steady state economy. Such a shift is similar to the gradual switch from a mass to an informative economy which Paul Hawken advocates as having the greatest potential in view of the current world economic situation.[20]

Drawing on the data gathered by the Worldwatch Institute, Lester Brown maintains that all the actions prerequisite to a changed world are already underway in some form, somewhere in the world. Several industrial countries have stabilized their populations. The birth rate *is* falling in a number of Third World countries. Many countries have adopted solar and other renewable energy policies. Industrialized nations since 1978 have charted a marked increase in the efficiency of their energy use. He writes:

> Each new hydroelectric generator, each new decline in the national birth rate, each new community garden, brings humanity closer to a sustainable society. Collectively, millions of small initiatives will bring forth a society that can endure. At first the changes are slow, but they

are cumulative and they are accelerating. Mutually reinforcing trends may move us toward a sustainable society much more quickly than now seems likely.[21]

Juxtaposed against the background of the daily newscast, a statement like this, however well documented, can sound like whistling in the dark. But all around us are signs both of the present order not working and a new one beginning to be forged. In the election of 1980, less than fifty percent of the population of the United States went to the polls. This was the lowest voter participation of any western democracy. The fact is a possible indication that many people are feeling emotionally, if not legally, disenfranchised. The enormity and remoteness of federal and state bureaucracies and the abstractions of national politics seem increasingly less connected to peoples' lives. This disassociation has a parallel development in the concomitant resurgence of activist grass roots political life. Thinking globally and acting locally has become more than a slogan of environmentalists and futurists. Falmouth's endorsement of the nuclear freeze, support of the Cape and Islands Self-Reliance Corporation Cooperative, and gathering interest in Cape Cod as a bioregion is our own local evidence. On a wider scale, we are in the midst of a country-wide proliferation of community-, county-, and neighborhood-based cooperatives, credit unions, recycling centers, health care and wholistic health centers, small businesses, and organizations for local food and energy independence. New vitality and commitment is being infused into neighborhoods, small towns, and rural areas, as people realize they can work effectively on smaller scales in the midst of huge, impersonal governmental structures. Evidence is mounting of renewed sensitivity to the local bioregion, as well as to local culture and traditions which bring out the identity of places and individuals. These are usually more closely tied to the immediate needs of people and environment. In small, diverse units throughout the country people are beginning to recreate the economic and political infrastructure for a more workable society. Giantism, in the form of the corporate state and an interdependent world mass economy seems to have reached its upper limits. Individuals and communities capable of faster and more direct responses than institutions, are beginning to react and to take steps to gain more control over the direction of their lives. Even though the message of quickening change has yet to be assimilated by the collective bureaucratic brain which, in the main, continues on its accustomed path, smaller units of evolution, more flexible and closer to the ground, are busy adapting to the world immediately around them. California writer Stephanie Mills, in discussing the phenomenon of the increasing strength of bioregional thinking said, "The resistance of people to colonization and cultural destruction is a constant of human history. A loyalty to the planet, a detailed sense of place and an informed love of homeland will save us."[22] Here at last perhaps is the reemergence of the instructions for want of which we have been so long adrift.

In the last quarter of the twentieth century the contrast has grown more apparent between the continuing militarism, competitiveness, corporate giantism, and exploitation embedded in a now-outmoded world view and the vision of human scale, ecologically sensitive, cooperative and peace-seeking entities of community and bioregion. Although the struggle is and will, no doubt, always be part of the human lot, from the limited vantage point of this moment in history, it appears that we are entering a period of intense polarization. Those who seek to keep open the greatest number of options for the unfolding of life on our planet have little force in the face of the vast weapons of planetary destruction arraigned against all of us. Yet even a mind of the stature of Gregory Bateson's did not discount completely the potential of an effort motivated by conviction. In *Mind and Nature* he wrote: "A fantasy is made real or validated by the actions that it dictates. In such a process the fantasy can then become morphogenetic and a determinant of society."[23] The drive for a nuclear freeze, the hope for a sane world, and for continued co-evolution with Gaia is based on considerably more than fantasy. A great deal of concrete, serious work is underway at every level, including that of the United Nations. In June, 1980, the Republic of Zaire presented to the United Nations a Draft World Charter for Nature which began:

> Life depends on the uninterrupted functioning of natural systems which ensure the supply of energy and nutrients. Life on earth is part of nature. Mankind has evolved from the same origins as other forms of life, and lives in constant interaction with them and the physical elements of the environment.[24]

The Greek myth of Pandora is a reminder that when, by dint of unrestrained human curiosity and meddling, all the evils had been let loose in the world, hope still remained in the bottom of the box. We still have hope. Often at New Alchemy we are asked why we keep plugging away, when the probable outcome for the future is so bleak. We can only say that in spite of our considerable intellectual pessimism, we have found in the work there, which is a small part of a much larger movement, a source of a kind of glandular optimism, evocative of what Greek mythology called hope. It was seen as a panacea in the face of the evils of the world—a transforming energy which can work, as though, in Doris Lessing's words, "there are lungs attached to men that lie as dormant as those of a babe in the womb and they are waiting for the solar wind to fill them like sails."[25] The work and the thinking that we have been describing are no longer dormant. They exist and struggle now toward a larger birth. We do not underestimate either the contest or the odds. This birth of a new way to order society represents an exercise in what our co-founder of New Alchemy, Bill McLarney, once defined as applied rather than theoretical love. At the Solar Village Conference Keith Crichlow remarked that apathy, not hate, is the opposite of love. We see strong signs of a quickening concern of applied love all over the world. The era of apathy, the era when humanity feels helpless to better its world, may be drawing towards a close. The predeominant image of our age remains, sadly, the burgeoning mushroom cloud hovering constantly over all of us. As the new cosmology emerges and begins to bind the scientific and the spiritual and to reinfuse all life with the essence of the sacred, that image could dissolve to one of the living, 'blue, true dream of Earth'[26] luminous in darkness of the surrounding sky. Closer to home one would catch glimpses of landscapes that are wild and landscapes that support ongoing generations of human and non-human communities, of villages and cities mirroring intense concentrations of activity and culture, of great wind-powered sailing ships on the seas and smaller ones plying the coasts and inland waters. It is a world not idyllically Utopian. It is still inevitably scrappy and difficult. But this is a planet and group of cultures determined to move toward an ever-unfolding potential, instead of gradual or total destruction.

It has been our intent with this book to paint, for our readers, word pictures of the possible—doing so in the frankly partisan hope that they will be persuaded of the validity of the concepts—an attempt on our part to say, as Robert Frost did, if less poetically, "You come too."[27] It will be a shared mutual venture into the unknown, offering for the present nothing more nor less than hope, in an ongoing search for those instructions which, if we begin to think and believe and act appropriately, may help us to go on living on our shining, blue-green home planet.

Notes and References

Chapter One

For further reading see:

The Book of the New Alchemists (New York: E.P. Dutton, 1977).

The Journals of the New Alchemists, Nos. 4-7. Available from:
    The New Alchemy Institute
    237 Hatchville Road
    East Falmouth, Massachusetts 02356

John Todd and Nancy Jack Todd, *Tomorrow Is Our Permanent Address* (New York: Harper & Row, 1980).

The New Alchemy Quarterly containing the most recent work of the Institute in research, outreach, and bioregional politics.

Chapter Two

1. Annie Dillard, *Teaching A Stone To Talk* (New York: Harper & Row, 1982), p. 15.

2. Carl Gustav Jung, *Memories, Dreams, Reflections,* recorded and edited by Aniela Jaffé (New York and London: Pantheon Books, 1963), pp. 143-144.

3. Giorgio de Santillana and Hertha von Dechend, *Hamlet's Mill: An Essay Investigating the Origins of Human Knowledge and Its Transmission Through Myth* (Boston: David R. Godine, 1977), p. 332.

4. Louis Mumford, *The Pentagon of Power,* Volume II of *The Myth of the Machine* (New York and London: Harcourt Brace Jovanovich, 1964), p. 86.

5. Mary Catherine Bateson, *Our Own Metaphor: A Personal Account of a Conference on the effects of Conscious Purpose on Human Adaptation* (New York: Alfred A. Knopf, 1972), p. 15.

6. Lewis Carroll, *Alice's Adventures in Wonderland* (New York: Modern Library).

7. Bateson, *Our Own Metaphor*, op. cit., p. 15.

8. Murray Bookchin, *The Ecology of Freedom: The Emergence and Dissolution of Hierarchy* (Palo Alto: Cheshire Books, 1982), p. 14

9. David Bohm quoted by Fritjof Capra, *The Tao of Physics: An Exploration of the Parallels Between Modern Physics and Eastern Mysticism* (Boulder: Shambhala, 1975), p. 138.

10. John Wheeler quoted by Capra, *The Tao of Physics*, op. cit., p. 141.

11. Charles Darwin quoted by William Irwin Thompson, *The Time Falling Bodies Take To Light* (New York: St. Martins Press), p. 93.

12. Gregory Bateson, *Mind and Nature: A Necessary Unity* (New York: E.P. Dutton, 1979), pp. 8-11.

13. Black Elk, *Black Elk Speaks: Being the Life Story of a Holy Man of the Oglala Sioux* as told through John G. Neihardt (New York: Pocket Books, 1972), p. 1.

Chapter Three

Precept One

1. James Lovelock, *Gaia: A New Look at Life on Earth* (Oxford, New York, Toronto, Melbourne: Oxford University Press, 1979).

2. Ibid, p. 9.

3. Ibid, p. 11.

4. Ibid, p. 152.

5. James Lovelock, "Daisy World: A Cybernetic Proof of the Gaia Hypothesis," in *CoEvolution Quarterly,* No. 38, Summer 1983, p. 71.

6. Stewart Brand introducing James Lovelock, "Daisy World: A Cybernetic Proof of the Gaia Hypothesis" in *CoEvolution Quarterly,* No. 38, Summer 1983, p. 66.

Precept Two

1. Frances Lappé and Joseph Collins, *Food First* (Boston: Houghton, Mifflin & Co., 1977).

2. E.F. Schumacher, *Small Is Beautiful: Economics As If People Mattered* (New York, Hagerstown, San Francisco, London: Perennial Library, Harper & Row, 1973).

Precept Four

1. Jim Dodge, "Living By Life: Some Bioregional Theory and Practice" in *CoEvolution Quarterly*, No. 32, Winter 1981, pp. 6-12.

2. Ibid.

3. William Blake quoted by E.E. Cummings, *Six Non-Lectures* (Cambridge: Harvard University Press, 1953), p. 32.

4. *The Register*, Yarmouth Port, Massachusetts, May 19, 1983, p. 16.

5. Arthur Palmer, *Toward Eden* (Ronkonkoma, New York: EBD Co.).

6. Dana Hornig quoting Arthur Palmer in *The Register*, May 12, 1983, p. 3.

Precept Five

1. Amory Lovins and Hunter Lovins, *Brittle Power* (Andover, Massachusetts: Brick House Publishing Co., 1982), p. 266.

2. Ibid, p. 369.

3. Jim Schefter, "New Harvest of Energy from Wind Farms" in *Popular Science*, Vol. 222, No. 1, January 1983, p. 59.

4. Ibid quoting Sandra Bodmer-Turner of U.S. Windpower, p. 61.

5. Lovins and Lovins, *Brittle Power*, p. 61.

6. Lester Brown, *Building A Sustainable Society* (New York, London: W.W. Norton & Company, A Worldwatch Institute Book, 1981,) p. 220.

8. William Becker, author of *The Making of a Solar Village* (Madison: The University of Wisconsin Extension Service) in personal communication.

9. Lovins and Lovins, *Brittle Power, p. 269.*

Precept Six

1. Louis Mumford, *The Pentagon of Power*, Volume II of *The Myth of the*

Machine (New York and London: Harcourt Brace Jovanovich, 1964), p. 155.

2. Gary Snyder, "Four Changes" in *Turtle Island* (New York: New Directions Publishing Corporation, 1969), p. 100.

3. Mumford, *The Pentagon of Power*, op. cit., p. 395.

Precept Eight

1. *Manas: A Journal of Independent Inquiry,* Vol. XXXII, No. 11, March, 1979.

Precept Nine

1. Gregory Bateson, "Form, Substance and Difference" in *Ecology and Consciousness*, edited by Richard Grossinger (Berkeley: North Atlantic Books, 1978), p. 30.

2. Papers by Keith Crichlow, Paul Sun and Malcolm Wells are included in *The Village As Solar Ecology: Proceedings of the New Alchemy/Threshold Generic Design Conference*, April 16-21, 1979. Available from
The New Alchemy Institute
237 Hatchville Road
East Falmouth, Massachusetts 02356

3. William Irwin Thompson, "On Food-Sharing, Communion, and Human Culture" from a sermon at The Cathedral Church of St. John The Divine, November 1, 1981.

4. Keith Crichlow, "Number and the Grail" in *The Lindisfarne Letter* No. 12 (West Stockbridge, Massachusetts: The Lindisfarne Press, 1981) p. 20.

5. Paoli Soleri, *The Omega Seed: An Eschatological Hypothesis* (Garden City, New York: Anchor Books, Anchor Press/Doubleday, 1981), p. 28.

6. "Arcosanti, Dream City" in *Newsweek*, August 16, 1976.

7. Soleri, *The Omega Seed*, op. cit., p. 165.

8. The Very Reverend James Parks Morton in *Heights: News and Events from the Cathedral of St. John The Divine*, Vol. 4, No. 2, Spring and Summer 1982.

9. David Bohm quoted by Fritjof Capra, *The Tao of Physics: An Exploration of the Parallels Between Modern Physics and Eastern Mysticism* (Boulder:

Shambhala, 1975), p. 138.

10. J.B. Priestley, *Literature and Western Man* (New York: Harper & Brothers, 1960), p. 445.

Chapter Four

1. da Vinci, Leonardo, *The Notebooks of Leonardo da Vinci*, edited by Pamela Taylor (New York, Toronto and London: New American Library, 1960), p. 186.

2. Theodor Schwenk, *Sensitive Chaos: The Creation of Flowing Forms in Water and Air* (London: Rudolf Steiner Press, 1965).

3. Jane Jacobs, *The Death and Life of Great American Cities* (New York: Random House, 1961).

4. Paul Hawken, *The Next Economy* (New York: Holt, Rinehart and Winston, 1983).

5. Bernard Rudolfsky, *Architecture Without Architects* (New York: Doubleday & Co., 1964).

6. Hassan Fathy, *Architecture for the Poor* (Chicago and London: The University of Chicago Press, 1973).

7. Jane Jacobs, *The Economy of Cities* (New York: Random House, 1969).

8. Paul Hawken, *The Next Economy.*

9. Personal communication at an international conference on biological agriculture in Zurich, Switzerland, May 1979.

10. John Hess in *The New York Times*, September 6, 1973, p. 32.

11. Claude Lévi-Strauss, *The Savage Mind* (Chicago and London: The University of Chicago Press, 1969).

12. Christopher Swan, "Light Rail" in *CoEvolution Quarterly*, No. 25, Winter 1981. Suntrain Inc. is located at 1717 Green Street, San Francisco, California 94123.

13. Wendell Berry, *Below in A Part* (San Francisco: North Point Press, 1980), p. 21.

Chapter Five

1. Henry T. Lewis, "Indian Fires of Spring" in *Natural History*, Vol. 89, No. 1, January 1980, New York, pp. 76-83.

2. Richard Critchfield, *Villages* (Garden City, New York: Anchor Press/ Doubleday, 1981).

3. Geoffrey Barraclough, Editor, *The Times Atlas of World History* (Mapelwood, New Jersey, and London: Hammond, 1978), p. 40.

4. Raymond William, *The Country and the City* (New York and London: Oxford University Press, 1973).

5. The Cornucopia Project of Rodale Press, *Empty Breadbasket* (Emmaus, Pennsylvania: Rodale Press, 1981.

6. Peter Raven and Helena Curtis, *The Biology of Plants* (New York: Worth Publishers, 1970), p. 681.

7. Eugene P. Odum, *Fundamentals of Ecology*, Third Edition (Philadelphia, London, Toronto: W.B. Saunders, 1971), p. 251.

8. John Jeavons, *How To Grow More Vegetables*, Second Edition (Berkeley, California: Ten Speed Press, 1979).

9. William McLarney and Jeffrey Parkin, *The New Alchemy Back Yard Fish Farm Book* (Andover, Massachusetts: Brick House Publishing, 1981).

10. Personal communication with Dr. Wes Jackson, 1983.

11. J. Russell Smith, Tree Crops: *A Permanent Agriculture* (Old Greenwich, Connecticut: Devin-Adair Co., 1953).

12. Bill Mollison, *Permaculture Two* (Stanley, Tasmania: Tagari Books, 1979), p. 1.

13. Gary Snyder, *Turtle Island* (New York: New Directions Books, 1974), p. 100.

14. Paul Hawken, *The Next Economy* (New York: Holt, Rinehart and Winston, 1983).

15. Daniel Deudney and Christopher Flavin, *Renewable Energy: The Power to Choose* (New York and London: W.H. Norton & Company, 1983), pp. 304 & 305.

Chapter Six

1. Donella Meadows et al., *The Limits to Growth* (New York: Universe Books, 1972).

2. Gerald Barney, *The Global 2000 Report to the President* (Washington: U.S. Government Printing Office, 1981).

3. Paul Hawken, James Ogilivy and Peter Schwartz, *Seven Tomorrows: Toward a Voluntary History* (Toronto, New York, London, Sydney: Bantam Books, 1982).

4. Jean Auell, *Clan of the Cave Bear* (New York: Crown, 1980).

5. Theodore Roszak, *Person/Planet: The Creative Disintegration of Society* (Garden City, New York: Anchor Press/Doubleday, 1978), p. 55.

6. James Lovelock, *Gaia: A New Look at Life on Earth* (Oxford, New York, Toronto, Melbourne: Oxford University Press, 1979), p. 147.

7. Daniel Deudney, "Whole Earth Security: Ageopolitics of Peace," Worldwatch Paper 55 (Washington: Worldwatch Institute, 1983), p. 20.

8. Steve Baer, "The Bomb" in *CoEvolution Quarterly*, No. 10, Summer 1976, p. 79.

9. Jonathan Schell, *The Fate of the Earth* (New York: Avon Books, 1982), p. 93.

10. Louis Mumford, *The Pentagon of Power*, Volume II of *The Myth of the Machine* (New York and London: Harcourt Brace Jovanovich, 1964), p. 81.

11. George Kennan, "On Nuclear War" in *The New York Review of Books*, Volume XXVIII, Numbers 21 and 22, January 21, 1982, p. 21.

12. Marie-Louise von Franz, C.G. Jung, *His Myth In Our Time* (New York: G.P. Putnam's Sons, 1975), p. 265.

13. Franz Kafka, "The Burrow" in *The Complete Stories of Franz Kafka* (New York: Shocken Books, 1946), pp. 325-359.

14. Schell, *The Fate of the Earth*, op. cit., p. 162.

15. *The Enterprise*, Falmouth, Massachusetts, February 26, 1982, p. 4.

16. Amelie Scheltema as quoted in *The Enterprise*, April 9, 1982, p. 14.

17. Gary Snyder, "Four Changes" in *Turtle Island* (New York: New Directions, 1969), p. 101.

18. Lester Brown, *Building A Sustainable Society* (New York, London: W.W. Norton & Company, 1981).

19. Hawken, *The Next Economy,* op. cit.

20. Brown, *Building A Sustainable Society*, p. 371.

21. Stephanie Mills, "Planetary Passions: A Reverent Anarchy" in *CoEvolution Quarterly*, No. 32, Winter 1981, p. 5.

22. Gregory Bateson, *Mind and Nature: A Necessary Unity* (New York: E.P. Dutton, 1979), p. 140.

23. Robert Frost, "The Pasture" in *Poetry and Prose* (New York, Chicago, San Francisco: Holt, Rinehart and Winston, 1972), p. 16.

We use the generic terms man and mankind as existing terminology or direct quotes, and with reluctance. Sexist language is one of the most obvious and revealing indications of the subordinate or secondary position women hold in the prevailing world view. We have taken care to avoid sexist language elsewhere.

Appendices

The Company of Stewards, Inc.
and Ocean Arks International:
A Joint Prospectus for Land Restoration
Prior to Agriculture and Settlement

A Project to Employ Advanced Ecological Concepts to Restore and Re-Inhabit Semi-Arid Mediterranean Coastal Regions

> Forests preceed civilizations,
> Deserts follow them
> Chateaubriand

Restoration Strategies: A Brief Overview

Orthodox reclamation projects involve tree planting and, in some cases, the transport of water to nurture young trees. Most reforestation schemes fail because they do not employ broad ecological strategies. We believe that receptive ecosystems must be created and put in place before rapid restoration and reforestation can be accomplished. This would include water, soil, organic matter and myriads of microorganisms, all of which might be absent at the outset. It should be mentioned that some of our ecological methods will remain proprietory until the first projects are successfully completed. This is not to lock up the information, but more to develop the initial credibility and economic strength of The Company of Stewards. Eventually, we want all of our techniques and biological materials to become available for comparable work by others. In fact, an advisory and supply service might well grow out of the schemes described here.

1: Topographic Restructuring

Most land forms are the result of weathering or erosive forces. Consequently, they lose water, soils and organisms, and they are incapable of catching and holding air- and water-borne organic materials, spores and seeds. Our first task on any site will be to restructure topographically key parts of the land. As a result the land will become the ecological equivalent of a baseball mitt catching and holding moisture, dust, soil and air-borne materials. Site design, requiring the one-time use of heavy machinery, is one primary ecological tactic we will use. Only a small percentage of the a-creage will require this treatment. Natural analogs will be developed on the remaining lands.

2: Desalination of Sea Water

In order for the land to be inexpensive, we assume that little or no water will be available on the site at the outset. Over a dozen water strategies will result in a rapid ecological development of the site. Two are mentioned here to illustrate ways in which The Company of Stewards uses salt water as a restoring agent.

i. **Salt Marsh Strategy:** Salt marshes exist at sea level. They are flooded by tides. Their biological roles are extremely important for the land as well as the sea. Salt water becomes fresh water in the tissues of its resident organisms, especially the plants. Salt marshes are nurseries for valuable marine life.

We propose to make "artificial" salt marshes above sea level and inland. They will be periodically flooded using New Alchemy windmill pumps and solar electric pumping units. The pumping will simulate tidal action. Salt marsh organisms will be collected throughout the Mediterranean region to seed and innoculate the new salt marshes.

In short order sediments will accumulate and new biomass will grow in abundance. Sediments and biomass will be used to mulch and "irrigate" adjacent terrestrial vegetation. Also, the salt marshes will capture rains and over time become increasingly brackish, ultimately supporting oyster and mullet culture. Salt-tolerant trees and crops, such as the Carob bean tree, will be established and will ring the inland salt marshes. These, in turn, will provide an ecological beachhead for other trees and organisms. Like the ripples of a pebble striking a pond, the soil and plant life will develop outward and inland from their marsh center.

ii. **Bioshelter Strategy for Water Use:** The second example employs the climatic envelopes, like New Alchemy's bioshelters, to desalinate sea water passively within. Trees, plants and soils, as well as fresh-water and marine fishes will be cultivated. The salt is separated out from the water by condensation along the roof of the bioshelter and on the sides of the translucent sea-water tanks. Pure salt is a by-product of this evaporative/condensation process.

The bioshelters will serve another critical function. They will be arranged spatially to provide the basis for an agricultural forest and its ecological components. The process is as follows:

a. A number of fifty- to eighty-foot diameter geodesic climatic envelopes, or bioshelters, are placed in a ring to create a large circle.
b. Sea water is pumped into solar silos within the bioshelters.
c. Evaporation and condensation produces fresh water.
d. Fresh water is used to establish a young ecological forest and its components *within* the bioshelter.
e. Excess water is used to establish a ring of trees immediately around the outer periphery of the structures.
f. Once trees, their root systems and soils are established (within two years), the geodesic bioshelters are lifted off and taken to a new site to repeat the cycle.
g. The cluster of trees and vegetation eventually trap their own moisture and trigger ecological succession preparing the way for a permanent agriculture.

Saline, brackish and fresh water all will have their unique ecological pathways. Aquaculture, the farming of aquatic organisms, will be integral to the restoration process. The abundant organic wastes produced by shellfish and fish will be used to help build soils. Cultured fishes and shellfish will provide an early economic base for the enterprise.

3: Soils

Weathering and microorganisms wear down rocks to produce soils. Some thirty-eight major groups of prokaryotic organisms, particularly bacteria, and higher eukaryota, including algae, molds, fungi, minute animals and vascular plants, do the work in concert. Inoculants of these living organisms combined with moisture and organic matter should allow us to produce the equivalent of a thousand years of topsoil in a decade. The secret lies in the combining of microorganism seeding with protecting newly forming soils from erosion and wind while providing them with essential moisture. Trace minerals for the soils will be derived primarily from salt marsh by-products. Initially, in some areas, it may be necessary to add missing macronutrients.

4: Trees

At the dawn of Mediterranean civilization most of the land was forested. We intend to recreate a forest-like landscape using a combination of economic trees, including fruit, fodder and nut-bearing species, with trees essential for the ecological unfolding of the orchards and soils.

The understory below the trees will be planted to various forage plants, including legumes, which have the ability to locate and bring trapped moisture up to the soil surface. Poultry, and eventually livestock, will be introduced at the appropriate stage to complete the nutrient cycles and initiate new economic activity.

5: Bioshelters

Bioshelters have already been mentioned under water and its use. Here it is suffice to say that a bioshelter is a structure with a solar envelope that is used to initiate a wide range of biological systems within. We have used such structures to grow food year-around in rugged temperate climates like the Maritime provinces of Canada. They are also used to grow vegetables and fishes in deserts and arid regions.

Recently, John Todd received support from the National Endowment for the Arts and two private foundations to design and develop a second generation bioshelter that would be long-lived, light and portable, capable of withstanding severe storms, exclusively solar-powered, adapted to extreme environments, including the Mediterranean, and cost-effective for food culture in the U.S.A. In this project these bioshelters will be the "embryos" and epicenters of the overall restoration project. Highly water-conserving, they will be used for plant propagation, for the hatching and culture of young fishes, and for the growing of myriads of beneficial microorganisms. Without them a ten-year time frame for each project would not be possible.

6: Wind and Solar Technologies

The Mediterranean is blessed with abundant sun and strong seasonal winds. Advanced wind engines and solar cell electronics, with which we are already experienced, will be employed for pumping, irrigating and electrical generation. New Alchemy's water-pumping windmills are already more cost-effective than electrical or diesel water pumping in the United States. Apart from liquid fuels for earth-moving machinery and overland transport, we do not intend to need petroleum products in large quantities. Each project will be energy independent to a high degree and invulnerable to supply disruptions or major fuel price increases. It should be emphasized that it is our careful linking of the technical, ecological, energy and agricultural elements into a unified approach that provides the basis for this relatively autonomous method of land restoration.

**A Request for Support
from Ocean Arks International**

A Project to Help Save the Coastal Fisheries of Guyana, South America, with the Aid of Advanced Sail-Powered Fishing Vessels.

I. Background and Purpose

The fisheries of many tropical nations are threatened by their country's international indebtedness, and as a result, the future of fishing communities is in jeopardy. Within the past three or four decades, fisherman have become totally dependent upon imported fuel, engines, boats, and spare parts/repair infrastructures, all of which were based upon access to foreign exchange. In many countries this critical hard currency is no longer readily available to fishermen. For example, Guyana, which has extraordinary fishery resources within its territorial waters, is particularly hard hit with a large foreign debt and soft currency. The artisanal fishermen find that fuel is becoming prohibitively expensive and engine spare parts are almost impossible to acquire. Collapse of the system is currently prevented by international aid organizations, such as the Canadian International Development Agency, making new motors available to the fishing cooperatives. This stop-gap measure is not expected to continue for much longer.

The International Monetary Fund has recently proposed another major devaluation of the Guyanese dollar which would limit even further their access to imported fuels and materials necessary to maintain their fleets. In theory the artisanal or coastal fishery could provide all of Guyana's protein needs, however, at the present time its marine resources are underutilized and many fishermen are nearly destitute. Unless a different approach to fisheries development is found, the situation can only worsen.

Guyana's problem is mirrored in various forms in many fishing countries, and the solution may be found in a traditional energy source.

The northeast trades blow year-round along the northeast corner of South America where Guyana is situated. Because wind power represents a genuine alternative for many regions, Ocean Arks International launched in 1979 a broad, multidisciplinary project to develop advanced design wind-powered fishing vessels.

Several design objectives were established at the outset. They were:

1. Components that required imported materials, and foreign exchange would be limited to 15% or less of the total cost of the vessels.
2. New wood-product technologies had to be found. They needed to be based upon easily planted and rapidly grown tree crops, and not on slow-growing, non-renewable noble woods which played a role in traditional boat construction and are now in short supply.
3. That the most advanced naval architectural and wind technologies be employed to create sail-powered working water craft capable of maintaining speeds equivalent to the motorized vessels they were intended to replace.

We felt that if the most modern engineering and materials science were applied to the challenge, then a new age of sail could come to the aid of fishermen. Further, we hoped to accomplish the task without adding to the ecological deterioration of the remaining tropical forests. Within the last year we have begun to achieve our technological and ecological objectives.

II. Accomplishments to Date

i. Wood-based Technologies: Three comparatively new technologies have been refined and adapted in order to build vessels with excellent strength to weight ratios and cost effectiveness. This includes Constant Camber mold utilization which allows identically shaped, mass-producible wood strips to be fabricated into compound curved hulls. Secondly, epoxies and wood veneers have been combined to create composite building materials which are strong, light, rot-resistant, and long-lived. Finally, vacuum bagging, a method for holding the strips of wood on the mold and providing pressure for the bonding stage in the fabrication of the wood/epoxy composite boat material, has been developed to the stage where it can be employed inexpensively in small communities throughout the Third World.

ii. Boat Wood Forestry: In collaboration with NAISA in Costa Rica we have established plantings of various trees to test as candidates for wood/epoxy composite construction. Melina, a fast-growing tree from New Guinea, has grown to a suitable size for boat construction within two years. Preliminary tests indicate it is epoxy compatible. Baromalli, an underutilized species from Guyana, has been successfully used to make the composite building materials.

iii. Naval Architecture: A prototype vessel, the first of the "Ocean Pickups," was launched in November of 1982. It was designed by Richard Newick, the creator of some of the world's fastest ocean-racing multi-hulls including the last OSTAR (Observer Singlehanded Trans Atlantic Race) winner "Moxie." He has successfully translated his skills into a working water craft. The prototype Ocean Pickup, a 32'-long trimaran which weighs less than a ton, has proven to be capable of carrying a ton and one half of cargo and of sailing at speeds up to 20 mph.

iv. Sea Trials and Fishing Gear Development: During the 1982/83 winter the prototype Ocean Pickup, the "Edith Muma," was given sea trials under a variety of weather conditions in New England. Trawling, trolling, gill netting, and long lining gear was developed for the boat and tested under difficult winter conditions.

v. Voyage to Guyana: During March and April 1983 the rig was modified, and the vessel was given a cabin and outfitted for the approximately 3,600-mile trip to South America. In May during the first leg, Cape Cod to North Carolina, and the second leg, Beaufort to Bermuda, the Ocean Pickup was continuously exposed to heavy-weather sailing and proved rugged and capable of handling the difficult conditions. The third leg of the journey, the 2,200 miles from Bermuda to Guyana against southeast headwinds, took just under fourteen days. This fast passage was made with the vessel carrying fishing gear, fuel, water, an auxiliary 15 hp outboard, supplies for Guyana, and provisions, attesting to its performance under load carrying conditions.

vi. Fishing in Guyana's Drift Net Fishery: In June and early July the Ocean Pickup fished in the drift gill net fishery of Guyana. Drift netters account for 40% of the artisanal catch. Because of their range and power requirements they are especially vulnerable to input cost increases. The Ocean Pickup successfully set, fished, and hauled one-mile-long gill nets under sea conditions considered unsuitable for existing boats. Fishermen had no difficulty sailing the multihull and were awed by its speed, effectiveness as a working platform, and its potential range. They told us they would willingly buy Ocean Pickups at our estimated in-country construction cost of G $35,000.

Our preliminary economic estimate indicates that each Ocean Pickup would save the fishermen more than G $20,000 annually and at least double their range. Based upon our rate of fish capture to date, Ocean Pickups in the drift net fishery could pay for themselves within a single year. The challenge now will be to create the infrastructure in order to cost-effectively build a fleet of Ocean Pickups in Guyana.

III. Project Work Still to be Undertaken

i. Experimental Fishing: A large mid-water zone exists between 5 fathoms (12 miles offshore) and 20 fathoms (35 miles offshore) which is beyond the range of most of the existing vessels and which is scarcely fished. We want to use our specially designed shrimp and bottom fish trawls for experimental fishing in this zone.

The Ocean Pickup is also rigged to fish for large sharks and to aid with developing a shark fishery. Shark meat and fins are in high demand. It is planned to fish in areas where sharks are considered abundant.

A single test trip using the trolling gear turned up catches of yellow fin, tuna, and mackerel. It would be valuable if we could establish a sail-powered trolling fishery for pelagic species.

ii. By-Catch from Shrimp Boats: Shrimpers trawl out to the territorial limits of Guyana in search of highly valuable shrimp. Up to 80% of their catch is fin fish which are discarded because they are not valuable enough to process on the trawlers which cost up to 1.5 million Guyanan dollars to purchase. Scientists estimate that almost all of the country's meat needs could be met by the fish which are killed and wasted at sea. As yet no one has found a cost effective solution to the by-catch problem.

We suspect an Ocean-Pickup-like vessel, larger than the prototype, may be the answer. To test our assumptions we propose to rendezvous with shrimpers at sea and transfer the by-catch to the Ocean Pickup, and with the fish packed in ice, sail the vessel the 70-90 miles to Georgetown. We have designed the transfer gear and are working on the logistics of rendezvousing with trawlers. We expect that within six months the cost effectiveness and suitability of Ocean Pickups as by-catch vessels can be determined.

iii. Developing and Financing the Infrastructure to Build Ocean

Pickups: We are committed to seeing that the preliminary research and development is translated into boat building and boat plantation enterprises. Over the next half year the various national and international agencies responsible for fisheries development will need to be coordinated and the entrepreneurial skills located to undertake the task. Ocean Arks International will coordinate the negotiations and seek financing for the implementation phase. Preliminary talks with the Inter-American Development Bank and Guyana Fisheries Limited have begun. Over the longer term OAI will have to provide ongoing technical and training services to the Guyanese in order to ensure that the finest possible fishing vessels are developed and made available to the people of the southern Caribbean region.

The Ocean Pickup in Guyana

In early May 1983 the Ocean Pickup, our thirty-two foot trimaran fishing vessel sailed from Martha's Vineyard in New England to Guyana in South America, a voyage of over three thousand five hundred miles (5,632 km). At the request of the Guyanese government and with support from the Canadian International Development Agency, we were to introduce our sailing technology to Guyanese fishermen.

The background leading up to the voyage was given in the first issue of Annals (Vol. 1, #1, 1583) with a detailed description of the Ocean Pickup and the technological innovation involved in its development. In this article therefore I will only summarize this briefly. Copies of the first issue of Annals are still available. I originally became involved with sail-powered working watercraft after direct experiences with some of the crucial problems being faced by fishermen throughout the tropics. Within the last few years, one fishing community after another has begun to suffer from a lack of engine spare parts and from the high cost of fuel which is often in short supply. Modern fishing vessels are getting harder and harder to maintain. In Guyana some fishermen have to own five outboard engines to keep one running and in spare parts.

It is not likely to get better. Most fishing nations started to modernize in the decades between the 1950s and the present. They obtained fast boats and efficient gear, and most of them caught enough more fish to pay for the changeover. Modernization made them dependent upon industrial nations for boats, gear, and fuels. To complicate the story this modernization was based upon international borrowing and national solvency which was removed and independent of the fishing communities. Now many of the tropical countries are becoming broke, to all intents and purposes, due to a lack of foreign exchange. Without hard currencies they are unable to import, and as a result, the service networks as well as the industrial infrastructure of the fisheries are beginning to fall apart—in some cases rapidly. Around the world there are growing numbers of small scale fishermen who now lack the where-with-all to ply their fishing trade.

It seemed to me that there must be a contemporary alternative to buying boats, engines, fuel, and gear from industrial countries. I started to look for more regional solutions that borrowed from the fruits of scientific and engineering knowledge which could be applied in the context of tropical countries and peoples. From the outset I set four basic guidelines or objectives for a project to help fishermen with the development of a new type of working vessel. They were: our fishing boat had to be primarily wind powered, but at the same time as fast as most of the motor boats it was to replace; construction technologies had to be suitable for building in the tropics, within the communities themselves; the primary construction material must be derived from fast growing trees which would be a part of the reforestation projects we intended to promote; finally imported components had to be less than twenty percent of the overall costs of the vessel. In this way, by exporting one in five of the vessels built into hard currency countries, input needs could be paid for.

These objectives would have been almost impossible to meet if it hadn't been for naval architect Richard C. Newick. Dick is known in yachting circles around the world for his record-breaking proas and trimarans which look like space age sailing craft and seem half bird and half boat. His major commitment is to use his design skills and technologies to create sailing workboats for fishermen, even the poorest of them. The first Ocean Pickup represents the marriage of three technologies: constant camber molding, wood/epoxy composite building materials, and vacuum bagging, with the advanced design of Dick Newick. The 32 foot trimaran is fast, weighs a ton, and can carry a ton and a half of gear and fish. It is strong, seaworthy, and we believe long-lived.

The first Ocean Pickup, the "Edith Muma," was fished off Cape Cod. In May skipper Russ Brown and Jonathan Todd sailed her down the east coast of the United States to Beaufort, North Carolina, and after making some alterations to the rudder, directly off-shore east to Bermuda. The trip was one storm or blow after another. Although it was an exhausting trip, the "Edith Muma" was beginning to prove herself at sea.

At the beginning of June I joined the boat in Bermuda for the approximately twenty-two-hundred-mile journey to Georgetown, Guyana. Russ stayed on as skipper while Jonathan flew to Guyana to help prepare for our arrival there. At dawn on the 3rd of June we set sail from Bermuda, and within a few days in the Sargasso Sea, we were plagued with light, fluky winds. We headed east toward Africa in the hope of picking up the trades. Our southeasterly course took us over four hundred miles east of Bermuda onto a line directly north of Georgetown, Guyana. Overall the voyage was fast and beautiful. The winds came mainly from the southeast and we were close hauled and drenched in spray on deck. The pocket-sized cabin could only hold one of us at a time. We alternated sleeping and sailing.

At 8:30 in the evening on the 7th of June I wrote in my log: "Sailing hard and fast, a trimaran at speed must rank as one of the penultimate sport sensations." By 10:30 my mood had changed. "I am worried that we are sailing too fast, for we are starting to emulate the little flying fish too closely." I doused the jib and eased the main. At 1:30 a.m. I collapsed into the bunk exhausted. On the 12th we were struck by a brief storm straight out of Caine Mutiny. We rode it out with a sea anchor and the rudder removed.

Our solace was that we were making distance over the water. By the end of the 12th day we reached our closest point to a body of land. We were 90 miles due east of Barbados. There was real beauty everywhere at sea. On June 12th I wrote: "The sky this dawn was beyond the painters or photographers art. It was a melange of massive, ever upwelling, ever dark and light permeated rainclouds. Quintessential tropical storm sky."

On June 13th: "This morning I fight off sea blindness by being the Darth Vadar of the sea. I have a face mask, visor, glasses, head band, foul weather top, and a 'Solar Aquaforms' t-shirt tied to my head, as the heat and light are so intense."

Inside the territorial waters of Guyana we started to see fishing boats, the first since Bermuda. Beyond the continental shelf we spotted a snapper/grouper long liner rolling the large seas. Later in the day we saw the first shrimp trawlers, working in pairs or clusters. During the night, as we closed in on the South American coast, more boats appeared. At one point the beacons I thought might be the Georgetown entrance turned into bobbing kerosene lights marking the extremities of huge fishing nets on the surface. During this last night at sea I had the distinct pleasure of slowly sailing past a freighter on the same heading. It gave me hope for an age of commercial sail.

At dawn on the 17th of June, after two weeks at sea, I found myself weaving through a phalanx of fishermen's traps, lines, and gill nets trying to hold the course Russ had set. As it turned out, his navigation through those current-filled waters had been flawless. The land in that part of the world is so low that, on a boat like the "Edith Muma," it is invisible eight miles offshore. Finally, I spotted the Pegasus Hotel, a landmark which indicated that we were on our course after two thousand miles. I woke Russ, and we called the lighthouse keeper on our hand-held radio. His voice was friendly, and he said he would telephone Robert Williams, the head of Guyana Fisheries Limited.

It was a different world as we entered the Demerara River. The "Edith Muma" sailed past the big market, close to an aging ferry, sardine-packed with people who stared at us, and then on up to the shrimp boat docks where we dropped sail. Waiting there to meet us were Neil Wray, the Guyanese coordinator of our project; Steven Drew, a master fisherman who was to work with us before taking up his teaching position in Fisheries at the University of Rhode Island; and Jonathan Todd and Rob Robinson, who were to crew during the fishing trials.

Without delay we began to ready the boat for fishing trials which would continue, except for an interruption for repairs and painting, until late August. Our first task was to include the Ocean Pickup in the existing fishery under the guidance of a local fisherman whose gear we used. The first fisherman, Henry Bosdeo, was a propitious choice. He lived in Zeeburg on the West Demerara coast and was one of the best at his trade. There was mutual respect from the beginning. Henry Bosdeo is a member of the artisanal fishery. The fishermen work with drift gill nets from open boats. The majority of the vessels are large, flat bottom skiffs up to 35′ (10.7 m) in length, powered by 40-55 hp outboards. Some of the boats have small cuddies forward to help protect the crew from the elements. Fishermen like Henry use mile long (1.6 km) gill nets to fish on the surface in the shallow waters of the continental shelf. Usually they stay within fifteen miles (24 km) of the shore. Henry Bosdeo and his neighbors burn a lot of fuel, up to $90 worth Guyana per day. Since they fish about 20 days a month, this works out to about $1,800 Guyana per month. In U.S. currency this represents an annual fuel bill for an artisanal fisherman of over $7,000. Many of the fishermen use auxiliary sails to reduce their costs. The skiffs do not carry ice which is not normally available. Men like Henry Bosdeo feel their range and duration at sea are constrained by lack of ice and by fuel costs.

The Guyanese have a small number of larger, diesel-powered drift gill netters with deck houses, crew quarters, and ice holds. These vessels range further afield and fish with large 1,200 lb (544 kg) 1½ mile long (2.4 km) nets. They carry ice and fish the productive middle grounds. Some long line for snapper 100 miles (160 km) offshore on the edge of the continental shelf.

Despite the fact that Neil, Jonathan, Rob and Steve were sick, our first day of fishing with Henry was a success. In a three-hour set they caught grey snapper, shad, ocean catfish, and one mackerel, which we sold in the

Georgetown Guyana market for $153 Guyana. Henry was so delighted with the boat that day that he relieved the ailing crew by sailing it smartly home. Once ashore he offered to buy the vessel. In subsequent trips with Henry our catch rate climbed to up to 300 lbs (136 kg) of marketable fish per hour. On one occasion we set and hauled the mile long gill net in conditions too rough for the rest of the fleet to leave the shore. The "Edith Muma" was starting to make friends.

Henry Bosdeo and other fishermen told us that they would modify their fishing if they had Ocean Pickups. They would sail further out to the richer, scarcely fished middle grounds. Following our example, they would also take up trolling. Steven Drew had rigged the Pickup so that we could troll four lines while traveling to and from the fishing grounds. Every time out we caught small numbers of the prized King Mackerel, *Scomberomorus cavalla*, and once we landed a yellowfin tuna, *Thunnus albacares*. With motor boats, trolling is often too fuel-consuming to justify economically, but fast trimarans, capable of sustained trolling speeds of 6-7 knots under sail alone, open up a new pelagic fishery for Guyana. In any case a number of the fishermen we worked with sensed the new potential inherent in fuel independence.

The following table is my preliminary attempt to assess the economic viability of a 1.5 ton Ocean Pickup built in Guyana and operated in the gill net fishery. The catch value is based on the average price we were paid for our fish but the figures do not include valuable pelagic species caught while trolling.

An Ocean Pickup could net an owner/skipper close to $12,000 U.S. a year. The present official exchange rate is $3 Guyana = $1 U.S. I suspect the calculations might be conservative as they are based upon catch rates of 200 lbs (90.7 kg) per day, and some of the better drift gill netters with comparable hold capacity catch an average of 300 lbs (136 kg) per day.

It is possible to view sail power in the drift net fishery in another way, namely by the fuel saved. Fuel savings alone would pay for an Ocean Pickup in the 10 fathom, 20 mile (30 klm) offshore fishery in as short a period as two years. In the middle ground fishery in depths of 20 fathoms annual fuel savings over a 55 hp outboard would be about $12,000 U.S. a year, which be close to the price of a Guyana-built Ocean Pickup.

Under Steve Dewey's direction we managed to experiment a bit with our long long lining and shrimp and fish trawling gear, but there really was not time to debug the gear and prove much about trawling from an Ocean Pickup or long lining for sharks which is Steve's specialty. We did get a chance to test the "Edith Muma" against one of Guyana's biggest food loss problems, namely, the destruction of fishes in the shrimp trawling industry.

Guyana obtains urgently needed foreign exchange by selling valuable shrimp to the American market. Most of the shrimping is done by U.S. companies. Shrimping vessels usually catch fish as well. Fish form up to eight percent of the total catch. Because shrimp processing is expensive, the shrimpers cannot afford to take up time and space with less expensive fish. The fish killed in the trawl are usually thrown overboard. This fish-by-catch, as it is called, if it could be returned to shore, would be an important local food source. All attempts to find cost effective ways of retrieving the fish have failed, so far, although the government insists that the trawlers keep a percentage of their fish catch and return it to Georgetown.

A year ago an international fishery consultant had suggested that the Ocean Pickup might help solve the by-catch problem. We decided to run an experiment and rendezvous with a trawler at sea, transfer the fish, and return to the processing plant. The odds were long against a rendezvous taking place at all. We organized to meet with a number of vessels including Captain Robb of the shrimper "Weremsha." Without a radio direction finder, or a single side band radio to communicate with, it was a pig-and-a-poke task to find a boat in that expanse of sea. My dead reckoning put us in the predetermined area, and after several hours of circling we made contact with the "Weremsha." Our first attempt to rendezvous and transfer in a rolling sea was slightly hair raising. As the shrimp trawl, filled almost exclusively with fish, was hauled onto the shrimper, we doused sail and motored slowly alongside. The motion of the troller was different to ours, and the transfer procedure potentially dangerous until we had the idea of tying up, not alongside the trawler, but to the bridle which hung from the trawl doors at the end of the boom. Being tied this way freed us of the hazard of being held right against a heavily rolling boat which dwarfed the "Edith Muma." Jonathan and Rob jumped aboard the trawler and helped load and transfer eight boxes of fish. The whole task took us an hour. Filled with good cheer, we set sail for Georgetown. The trip back was a flying journey with the winds and tide in our favor. Four hours later we were tied up at the fish processing plant on the Demerara River.

The 1.5 ton Ocean Pickup may be too small to be optimal as a by-catch boat, and the "Edith Muma" lacks an insulated ice hold necessary to ensure the return of high quality fish. Dick Newick has designed a big sister to the "Edith Muma" which can carry up to three tons of ice and fish in its insulated hold. The 3 ton Ocean Pickup, if it were to be used as a by-catch boat, would have the communication and navigation gear for easy rendez-vous at sea.

A sail-powered by-catch boat in Guyana would be cost effective if the difference between the price of fish paid to the shrimp trawler captains and the price received for the fish were at least seventeen cents a pound, provided the vessel made one hundred trips a year. My calculations did not take into account fish caught while trolling to and from shrimping grounds.

The 3 ton Ocean Pickup could also be a very adaptable vessel. Like the smaller boat, it could be employed as a multi-purpose fishing vessel capable of drift gill netting, troling, trawling, and long lining. It would be easy to build the Ocean Pickup in two sizes. They could be made from panels that came off the same master mold.

Since the summer, various people from the Guyanese government, international development agencies, and the private sector have begun to assemble the infrastructure to build a fleet. Ocean Arks International would provide design and training assistance. Robert Williams, the executive director of Guyana Fisheries Limited, told the press he would like to see the project build at least two hundred vessels. The wheels of government anywhere usually move slowly, and Guyana is no exception. It is my hope that within a year or two the fisheries of the region will begin to change and that the Ocean Pickup will be a common sight on Guyana's fertile seas.

As we go to press, Dick Newick and I are shortly to sail the "Edith Muma" to Trinidad and Tobago and then along the Spanish main to Costa Rica. There we will join Bill McLarney and his NAISA colleagues on the Talamanca coast. They have already planted groves of potential boat-wood trees, Albizia, Sesbania, Eucalyptus, and Melina. The Melina has grown to boat-wood size in less than three years. Preliminary tests indicate it to be compatible with our construction technologies.

Some of the fishermen there are looking forward to the Ocean Pickup trials. Many of them along the Atlantic coast can no longer afford to operate their outboard-powered vessels as inflation and fuel scarcities have eroded their economic base. Our overall plan, in collaborating with Bill McLarney and NAISA, is to create an integrated scheme to assist both farmers and fishermen of the region. The overall project involves fishery research, boat building, reforestation, and agricultural diversification.

NAISA has been working on the food issue for ten years in the region. Within the ecological framework of our work there, the Ocean Pickup will be one component, which we hope will diversify the local resource base.

Index

Index

Before 8:00 in the morning